DEMYSTIFICATION OF GENE

The People who made it Possible

– Author –

Hirendra Kumar Das

Ph.D. (Biochemistry) and D.Sc. (Biochemistry) Calcutta University.

ASTRAL

Publishing Authors Across the Globe

2025

Daya Publishing House®

A Division of

Astral International Pvt. Ltd.

New Delhi – 110 002

ISBN: 978-93-5919-995-5

Published by : **Daya Publishing House®**
A Division of
Astral International Pvt. Ltd.
– ISO 9001:2015 Certified Company –
4736/23, Ansari Road, Darya Ganj
New Delhi-110 002
Ph. 011-43549197, 8130496929
e-mail: info@astralint.com
Website: www.astralint.com

Printed at : **Replika Press Pvt Ltd.**

PREFACE

'Textbook of Biotechnology' (ISBN: 978-81-265-6404-0; ISBN: 978-81-265-8359-1 [ebk]), and 'Gene and its Engineering' (ISBN: 978-81-265-4928-3) were meant for students and teachers of Molecular Genetics, Molecular Biology and Biotechnology. 'Demystification of Gene, the People who made it Possible' has been written for those who are not familiar with the above mentioned disciplines. The Annexures at the end of some of the chapters are for people who want to delve deeper into the subject.

Most people have only a vague idea about gene to be something that controls the hereditary traits. Comparatively short parents would like their children to choose taller spouses. The parents expect, but do not know why exactly the chances are good that the grandchildren would not be as short, if the spouses of their children are taller. A pet shop must have proper pedigree record, otherwise the puppies would not fetch good price. The buyers would assume that if the pedigree tells about the purity of a breed, the breed that is well known for some particular qualities that they like, then the puppies are likely to develop the same qualities when they grow up. The buyers would not even care to think why a particular breed is likely to have the same qualities generation after generation. If a neighbour's farm has produced sweeter corns, seeds for his farm would have to be procured from the same source next year. The farmer might have never wondered as to what exactly is the nature of information present in the small seeds that guide the production of sweeter corns year after year. Some people have heard about that word DNA, but that is it. A lawyer while trying to portray his client's opponent as inherently bad, would tell the judge, "its in his DNA Sir". But the lawyer does not know how the bad person's DNA would have made the person bad. May be, the judge also does not know.

The process of demystification of gene got initiated around the middle of the nineteenth century and was completed by little beyond the middle of the twentieth century. The people who worked for the demystification of gene, had remarkable determination and scientific acumen, but like ordinary people some of them were very selfish not caring to acknowledge other's contribution, while some others were very generous and supportive of each other.

This book takes the reader through the concept of gene as the hereditary element and talks about all the scientists who contributed towards understanding of gene and how it functions. The book tells that gene comprises DNA, that messenger RNA carries the message from DNA for the synthesis of proteins, that proteins act as biological catalysts for chemical reactions occurring in living beings, and that these chemical reactions distinguish one individual from the other, thus completely demystifying gene.

I am thankful to Rabindra Narayan Kundu, a retired Applied Physicist and my college time friend, for investing long hours in going through the several versions of the manuscript and suggesting improvements and helping me in trying to ensure that this book also does not appear like a textbook.

Hirendra Kumar Das

REVIEWS

Demystification of Gene, the People who made it Possible serves as a detailed narrative of the pivotal milestones in genetics, beginning with early observations such as Mendel's Factor and Johannsen's Gene, and progressing through the discoveries of key figures like Friedrich Miescher and Oswald Avery. The book excels in its portrayal of lesser-known but equally influential figures such as Garrod and Berzelius. The book provides an insightful account of the race to discover the molecular structure of DNA. From Friedrich Griffith's observation of Diplococcus pneumoniae to Oswald Avery's ground-breaking demonstration that DNA was the transforming principle to pivotal figures like Rosalind Franklin, whose X-ray diffraction images provided critical evidence of DNA's helical structure, and finally the triumph of Watson and Crick in constructing the double helix model. The discussions on the Central Dogma of Molecular Biology and the role of messenger RNA as proposed by Jacob and Monod are well-explained, demonstrating the shift from gene theory to the biochemical processes that underlie inheritance. This section describes the work of pioneers like François Jacob, Sydney Brenner, and Francis Crick. Finally the author portrays the events that led to the cracking of the genetic code. From Marshall Nirenberg and Har Gobind Khorana's work and Paul Zamecnik's identification of the intermediate step in protein synthesis. The book expertly outlines how these discoveries changed the landscape of Biology, making it one of the most significant scientific achievements of the 20th century.

The author's decision to frame these advances through the lives and work of the scientists who made them possible, brings a personal and accessible dimension to what might otherwise be a complex and abstract field. The author has also explained some facts through annexures given with some chapters. This human-centered storytelling will surely resonate with students and teachers alike and also general readers. I also feel that this book offers inspiration. It reminds readers that many great discoveries were the result of years of persistence, collaboration, and sometimes serendipity. Prof Das's ability to synthesize complex scientific concepts into a narrative that is both educational and accessible is one of the book's standout qualities.

Professor Sudhir K. Sopory
Emeritus Senior Scientist
International Center for Genetic Engineering and Biotechnology, New Delhi
Former Vice-chancellor
Jawaharlal Nehru University, New Delhi

In the book, 'Demystification of Gene, the People who made it Possible', the author has brilliantly captured the journey to the discovery of Genes in finer details, where every little step taken by scientists in the run up to the final goal is described chronologically, covering a period of close to 4000 years. Author begins his long story in what should be comfortably a familiar territory for his readers, describing how, consciously or otherwise, scientists have meticulously worked on different forms of life over centuries, leading to the general understanding of hereditary biomolecules. He goes on to describe how the understanding of protein and nucleic acids evolved by intelligently joining the pieces of information available in a jigsaw puzzle in a bid to establish their molecular structures. This, he attributes, led to the discovery of DNA structure, and later on to the proof of the existence of Genes, as an universal descriptor of life phenomena. Concluding that those individuals that carry better Genes from the rest in some advantageous way will stand a better chance of surviving, and of passing on those variations to future generations. It is a sheer joy to read this interesting book, because of the lucid style of writing using simple English, a must read for the curious and the expert alike.

Professor Himadri B. Bohidar

Emeritus Professor, Delhi Technological University, Delhi

Founding Chairperson, Special Centre for Nanoscience,
Jawaharlal Nehru University, New Delhi

Former Professor and Dean, School of Physical Sciences,
Jawaharlal Nehru University, New Delhi

A unique feature of the book, 'Demystification of Gene, the People who made it Possible' is the historical description of events starting in 1894 B. C. in Babylon to what we are today. Knowing the names of the scientists and seeing them in pictures creates an entirely different impression of the events in the mind. The readers will be excited to know how young were these to-be scientists, when they perceived the ideas and designed experiments with facilities available to them. From palm trees bearing dates to Mendel's peas in the garden and cross- pollination experiments by Johannsen, that led to the terms "genotypes and phenotypes".

Enzymes being Biocatalysts , the nature of proteins and further discoveries describing mechanism and regulation are discussed in a sequential manner.

The elucidation of the structure of DNA, the " Central Dogma of Molecular Biology" followed by the present day of knowledge, is presented in a very lucid manner. Another feature of this book is the description of key experiments performed for most discoveries.

This book is ideal for all undergraduate students and provides a guideline for decision making in their future endeavours.

I recommend the book, 'Demystification of Gene, the People who made it Possible' for all undergraduates whether pursuing their studies in Arts or Science.

Professor Shail K. Sharma
Former Professor and Head, Department of Biochemistry
All India Institute of Medical Sciences, New Delhi

The book "Demystification of Gene, the People who made it Possible" written by Professor H. K. Das is a first of its kind, which delves into the historical milestones and ground breaking discoveries made by different researchers that ultimately led to the development of the Gene concept.

The story begins in 1894 B.C. in Babylon, where King Sumuabum expressed displeasure over the less sweet dates served to him, which motivated a wise old man in his court to resort to the practice of manual pollination of date palms, improving the sweetness and quality of the fruit (Chapter 1). This early example of selective breeding sets the stage for the book's exploration of genetic science.

From there, the story takes us to Old-Brünn (now in Czech Republic) where Gregor Johann Mendel's famous experiments on pea plants led to the formulation of the Laws of Heredity, which ultimately earned him the status of "Father of Modern Genetics" (Chapter 2). The story further continues to explain the readers how the subsequent developments in the fields of medicine, biochemistry and microbiology spearheaded by dedicated efforts of eminent personalities like Garrod, Berzelius, Pasteur, Kuhne, Buchner and Sumner, led to the understanding of the chemical reactions in living beings and discovery of the enzymes, their catalytic roles and their protein nature. These advancements laid a strong foundation for establishing the link between the enzymes and the hypothetical genes (Chapters 3-18).

One of the most compelling aspects of this book is its detailed exploration of the personal histories of the scientists. In each chapter, the author delves into the challenges these researchers faced, such as economic hardships and the impact of World War II, and how they pursued their scientific passions despite these obstacles. For example, George Beadle, whose father was a farmer, was initially expected to follow his father's footsteps, but thanks to the encouragement of his high school teacher, Bess MacDonald, Beadle pursued higher education. Similarly, the story of Erwin Chargaff's early struggles during Vienna's 1923 inflation and his eventual success in pursuing a PhD at Yale University, is a testament to his perseverance. The book is abundant with such inspiring stories, filled with personal and professional struggles. The efforts of the author in collecting as many details as possible related to the scientists are clearly evident and these stories serve as powerful reminders of the importance of determination and dedication in science, especially to the young readers who try to pursue their career in science.

Chapter 19 delves into the life and work of Rosalind Franklin, a brilliant scientist whose famous Photo 51 played a crucial role in elucidating the double helical structure of DNA. Despite the lack of recognition during her lifetime, her dedication to science is brilliantly portrayed. The chapter also covers the conflict she had with Maurice Wilkins over guiding a PhD student and her battle with cancer, which she fought while maintaining her commitment to science. The subsequent chapter explores the race for the discovery of DNA double helical structure focusing on Watson and Crick, who announced it on Feb 28, 1953 at the Eagle's Pub, Cambridge. The chapter explains very lucidly how the haste in publishing first or claiming credit, sometimes persuades the researchers to false practices. In order to win the race over Pauling in discovering the DNA structure, how Watson allegedly took the famous Photo 51 of Franklin without giving her due credit and how Wilkins helped him in doing so, are clearly explained by the author. The aggressive tactics of James Watson were also evident in another incident, when he stopped Jacob and Brenner to publish their results on messenger RNA until their group arrived at the results (Chapters 24-25).

The discovery of messenger RNA by Francois Jacob, Sydney Brenner and Mathew Meselson and also James Watson and co-workers shifted the focus of research from DNA to RNA. Chapter 26 describes the RNA tie club which consisted of 20 members representing 20 amino acids and each member wearing a tiepin with a three letter code of the amino acid that was allotted to them. Chapter 27 describes the ground breaking discovery of Marshall Nirenberg that messenger RNA is required for protein synthesis and polyU acts a messenger RNA for polyphenylalanine, leading to the identification of UUU as the code for phenylalanine. Nirenberg's discovery was described as the "biological Rosetta Stone" and had a profound impact on the scientific community. Subsequent chapters describe how the publication made by Nirenberg in 1961 inspired several researchers, including Har Gobind Khorana and Severo Ochoa, the non-member's of RNA tie club, to shift their focus to Molecular Genetics and how their contributions helped in deciphering the complete genetic code (Chapters 28-30).

The story concludes with the discovery of the adapter molecule, the transfer RNA, the protein synthesis and the central dogma of life through the contributions of Francis Crick, Paul Zamecnik, Robert Holley (Chapters 31-34).

In every chapter, the author weaves together the story of the gene with simple yet captivating language, making complex scientific ideas understandable and compelling. The book stands as a true tribute to the pioneering scientists who dedicated their lives to unlocking the mysteries of genetics through demystification of the gene. Their stories of perseverance and passion, so elegantly penned down by the author, would inspire all readers. I thoroughly enjoyed this enlightening and inspiring read and wish the author continued success in his future endeavours.

Professor Ashok K. Singh

Former Director, ICAR Indian Agricultural Research Institute
New Delhi

Assisted by Dr. Haritha Bollinedi, Senior Scientist, Division of Genetics
ICAR Indian Agricultural Research Institute, New Delhi

Demystification of Gene: The People Who Made It Possible by Hirendra Kumar Das is an engaging and informative introduction to genetics and molecular biology for non-biologists and those without much scientific background. The book takes readers on a historical journey through seminal discoveries in genetics and molecular biology, introduces the scientists behind these breakthroughs, and elucidates how genes function in shaping organisms and transmitting hereditary information.

The book excels in narrating the historical progression of discoveries in genetics and molecular biology. Starting from the ancient practices of selective breeding to the ground-breaking work of Gregor Mendel, the storytelling approach effectively captures the curiosity and persistence of various scientists involved in unravelling the mysteries of genes, genetics, and molecular biology.

Manual pollination was practised in 1894 B.C. in the date palm orchards of Babylon, so those people knew that the tiny pollens carried information that could influence the size and taste of the dates. For thousands of years, the practice continued, but nobody had any inkling about the nature of the information the pollens carried.

The field of genetics as we know it today started with Gregor Mendel's work on the inheritance of different traits in garden peas around 1856. Mendel studied the transmission of simple traits such as red vs white flowers, wrinkled vs round seeds tall vs short plants, etc. Mendel observed that for each trait only one type i.e. tall appeared in the immediate progeny whereas in the grandchildren both tall and short types were observed. Mendel also found that if he took two different traits together i.e. height and flower colour, they were inherited independently of one another. He inferred that independent hereditary "factors" were controlling the traits.

While today Mendel strides in the field of genetics as a colossal figure, his work was too novel for the understanding of the scientific community at that time. His work was ignored till at the beginning of the 20th century when three laboratories independently working with different plant species came to the same conclusions as Mendel had.

In 1902 Edward Garrod published a paper in which he implied that Mendel's hereditary "factors" worked by controlling chemical reactions in an individual, whether it was a plant or a human being. Garrod referred to the hereditary ailment named alkaptonuria in which the urine of a person turns black. Garrod ascribed this ailment to the absence of a chemical reaction due to the person inheriting a defective hereditary factor.

In 1909 Johannsen coined the term "gene" to describe the hereditary "factor" and the word has stuck on. The author takes us on a journey to various laboratories that contributed towards putting various pieces together to solve the jigsaw puzzle of what genes are, what they do and how. We go through the research efforts of Berzelius who rightly guessed that chemical reactions in living organisms might be mediated by catalysts. Wilhelm Kuhne coined the term "enzymes" for the biological catalysts. Sumner crystallised the enzyme urease and found that it was a protein. Earlier Hofmeister and Ficher had independently shown proteins to be linear arrays of amino acids. In the 1930s, the painstaking efforts of Beadle and Tatum created mutations in bread mould *Neurospora* correlated inactivated enzymes with specific genes that became defective due to mutations caused by X-irradiation.

What chemical entity constituted genes? Here the author rightly takes us to the laboratories of Griffith and then Avery, the latter clearly showing that purified DNA from an infectious bacterium *Diplococcus pneumoniae* transformed a non-infectious variant into an infectious form. Chargaff in 1950 found that of the four bases that constitute DNA, two purines adenine (A) and guanine (G) and two pyrimidines cytosine (C) and thymine (T), irrespective of the source of DNA, the proportion of A was the same as that of T and the proportion of G was the same as that of C.

In 1953, Watson and Crick published their landmark paper on the molecular structure of DNA. The author has rightly pointed out that Watson and Crick could not have reached the correct structure of DNA without the crucial clues from Rosalind Franklin's X-ray crystallographic work and calculations which were shown to Watson by Maurice Wilkins a colleague of Rosalind Fraklin without her knowledge. From this important milestone, the story takes us to how information is organized in the DNA and how it is converted into proteins. We now meet eminent scientists like Jacob and Monod who in 1961 showed that RNA (aptly named Messenger RNA) took the information from DNA to proteins. As there are 20 amino acids, Sydney Brenner and others predicted that an amino acid is encoded by three bases and proved it. Thereafter, the book explains in a simple way the work of Marshall Nirenberg, Har Gobind Khorana, and Severo Ochoa in cracking the genetic code.

Crick asserted in 1955 that structurally it was impossible for an amino acid to directly recognize ribonucleotides. He enunciated "the adaptor hypotheses". The adaptor molecules would contain complementary nucleotides which would recognize specific triplets in mRNA. Zamecnik showed the presence of transfer RNA (tRNA) which carries amino acids. Holley purified alanyl tRNA and sequenced it. He found a triplet that serves as an anticodon to triplet codon GCU of alanine. Subsequently, the base sequences of all other tRNAs were determined following Holley's methods.

The book excels in narrating the historical progression of the major discoveries in the area of genetics and basic molecular biology that elucidated what genes are, how proteins are made and contribute to traits that we can observe and how these traits are inherited. The author also provides meaningful insights into the lives of the scientists behind these discoveries and highlights the passion with which they pursued their interests. This humanizes scientific discovery and fosters a deeper connection with the subject. The chronological presentation of discoveries helps readers understand how each scientific breakthrough was built upon previous knowledge, creating a coherent narrative that logically progresses toward the modern understanding of genes.

Unlike many authors who focus on medical aspects to attract readers to basic genetics and molecular biology, Hirendra Kumar Das takes a more challenging path by introducing lay readers to the fundamentals of what a gene is and how it functions. This approach enriches the reader's foundational understanding and sets the stage for appreciating the broader implications of genetic science that are unravelling before us today.

This book put together by Prof. Hiren Das is a valuable resource for anyone curious about the science of heredity and molecular biology. I am sure those who read it will come out wiser on how scientific knowledge advances. Many young students of science will get inspired to take up solving challenging problems in science rather than opt for more secure professions. This book is therefore not only for those with non-science backgrounds but for everyone curious about the progress made in life sciences in the 20th century.

Professor Deepak Pental

SERB-National Science Chair, Centre for Genetic Manipulation of Crop Plants

University of Delhi South Campus

Former Vice-Chancellor, University of Delhi

CONTENTS

Preface v

Reviews viii

1. The wise old man of Babylon 1
2. Mendel's Factor 7
3. Johannsen's Gene 13
4. Garrod's study of Alkaptonuria, a disease that appeared hereditary 17
5. Berzelius's Catalysts 21
6. Pasteur's Vital Force 25
7. Kühne's Enzyme 29
8. Buchner's alcohol production without live yeasts 31
9. Sumner's discovery - enzymes are proteins 35
10. Mulder described the main constituent of albumin to be "protein" 41
11. Suggestion of Hofmeister and Fischer that proteins consist of many amino acids linked to each other 45
12. Beadle and Tatum's one gene one enzyme hypothesis 53
13. Frederick Griffith's observation on *Diplococcus pneumoniae* 61
14. Oswald Avery's demonstration that the transforming principle of Griffith was nothing but DNA 65

15. Friedrich Miescher had discovered DNA as early as 1869 69

16. What is the chemical nature of DNA? 71

17. Erwin Chargaff determined the amounts of different bases of DNA 75

18. In a bid to understand the three dimensional structure of DNA, Rosalind Franklin was successful in determining its X-ray diffraction pattern 79

19. James Watson and Francis Crick constructed a molecular model of the double helical DNA 85

20. Francis Crick, Sydney Brenner and co-workers demonstrated that three bases in DNA coded for one amino acid 99

21. Francis Crick's Central Dogma of Molecular Biology 107

22. What is RNA? 109

23. Observations of Brachet implicated involvement of RNA in protein synthesis 111

24. François Jacob and Jacques Monod mooted the concept of messenger RNA as carrier of message from DNA 113

25. Search for the hypothetical messenger RNA of Jacob and Monod 119

26. The RNA Tie Club 123

27. Marshall Nirenberg initiated cracking of the genetic code as a first step in understanding how proteins are made 125

28. Har Gobind Khorana also played a stellar role in cracking the genetic code 131

29. Severo Ochoa also played a role in cracking the genetic code 135

30. The following are the genetic codes for the different amino acids 139

31. Francis Crick's adaptor hypothesis to explain how the protein synthesizing machinery translated the bases of the triplet codes in the messenger RNA into the amino acids 141

32. Paul Zamecnik obtained the first clue of the intermediate step in protein synthesis for translating the triplet of bases in the messenger RNA into the amino acids 143

33. Robert William Holley determined the base sequence of a transfer RNA and revealed the presence of an anticode in it 147

34. Synthesis of proteins 153

35. Recapitulation of the story in brief 157

Source Locations *167*

1

THE WISE OLD MAN OF BABYLON

The story of Demystification of Gene began in 1894 B.C. in the court of the new king Sumuabum of the city of Babylon. The 3rd dynasty of Ur had fallen and Babylon, the prosperous city state on the bank of the river Euphrates, had been annexed by Sumuabum, the chief of the migrant Amorite tribe, who originally did not belong to Mesopotamia. Sumuabum was relishing the local dates, but found these to be much less sweet, though larger in size. He expressed his displeasure about the dates in the court and an old man who had come to see the new king, silently took up the challenge.

The old man remembered that one year all the date palm flowers were eaten up by rats and there were no dates that year. He had observed that all date palm trees did not bear fruits, though all date palm trees would flower. He also observed that the fruit bearing date palm trees would bear very few fruits, if there were no non-fruit bearing date palm trees in the vicinity. He had concluded that something from the flowers of the non-fruit bearing date palms, may be the pollens, were possibly being carried by the wind and was necessary for fruiting of the fruit bearing date palms. He likened the date palms that did not bear fruit to the human male.

The old man searched outside Babylon for date palm groves which produced sweeter dates (though smaller) and collected the flowers from their male plants. He dusted the pollens from the male plant flowers on to the flowers of female plants in Babylon. When the fruits appeared later, some of those larger dates were indeed sweeter. Since then it had been a regular practice in the date palm orchards of Babylon to manually pollinate.

The practice worked year after year, century after century, but nobody had an inkling as to what exactly the tiny pollens could be carrying from the male plant, that was not only essential for the flowers of the female plants to be converted into fruits, but was also contributing to the quality of the fruits. Figure 1 shows a date palm tree full of date palms, Figure 2 shows male date palm flowers, while Figure 3 shows female date palm flowers.

Legend of Figure 1. A date palm tree laden with date palms. *The photograph was kindly made available by Dr. Dhurindra Singh of the ICAR-Central Institute of Arid Horticulture, Bikaner, Rajasthan, India.*

Legend of Figure 2a. Photograph of a male date palm inflorescence. *Courtesy Dr. Pranab Kumar Mandal, ICAR National Research Institute for Plant Biotechnology, New Delhi, India.*

Legend of Figure 2b. Close up photograph of a male date palm inflorescence.
Courtesy Dr. Pranab Kumar Mandal, ICAR National Research Institute for Plant Biotechnology, New Delhi, India.

Legend of Figure 3a. Photograph of a female date palm inflorescence. *Courtesy Dr. Pranab Kumar Mandal, ICAR National Research Institute for Plant Biotechnology, New Delhi, India.*

Legend of Figure 3b. Close up photograph of a female date palm inflorescence.
Courtesy Dr. Pranab Kumar Mandal, ICAR National Research Institute for Plant Biotechnology, New Delhi, India.

2

MENDEL'S FACTOR

We wait for four thousand years for our story to advance and move from Babylon to Old-Brünn, which is now situated in the Czech Republic. It was 1856, a monk was manually pollinating pea flowers in the garden of St. Thomas-monastery. Why was the monk pollinating flowers? Growing peas was the job of the gardener. In addition to religious duties, a monk had to teach the new entrants in the Abbey. This particular monk, Gregor Johann Mendel, was a Physics teacher, but he nurtured an innate curiosity about heredity. He initiated experiments with mice to study heredity, but the other monks did not approve of experiments with animals having sex. So Mendel switched to peas.

Mendel, was born in 1822 in a German speaking poor farming family of Moravia, which is now situated in the Czech Republic. In his early life, Johann Mendel, that was his name then, worked with his two sisters in his father's farm which had been owned by the Mendel family for at least 130 years. Soon Johann's crave for knowledge made him leave the ancestral farm and join the University of Olomouc, where he studied Philosophy and Physics. Because of financial hardship, however, he soon had to discontinue studying in the University.

He then decided to join a monastery as he felt, "a monastery was the best place to advance education without having to pay for it". F. C. Napp, abbot of the St. Thomas-monastery of Old-Brünn, admitted him in his convent and Johann took the name Gregor. In the monastery, Napp helped him study natural science in general and metereology and bee keeping in particular. Subsequently Napp sponsored him for advanced studies in the University of Vienna.

After completing his studies in Vienna, Mendel returned to his abbey in 1853. Now he wanted to pursue his curiosity about heredity. He instinctively felt that the laws of heredity would be applicable the same way irrespective of whether it was mice or peas or date palms or humans. He probably chose peas and not any other plant like date palm, because peas would flower in a few months after sowing, while date palm would take several years. One advantage of the pea plants is that these are small enough for cultivating thousands of plants, but the pea flowers are big enough for dissection and removal of the stamen, the male organ, leaving the ovary, the female organ intact. Figure 4 is the photograph of a cluster of pea plants with couple of pea pods.

Legend of Figure 4. Photograph of a cluster of pea plants.

Legend of Figure 5a. Photograph of a white pea flower in the pea plant. The actual width of the flower is little more than 1 cm.

Figure 5a is a photograph of a white pea flower in a pea plant, while figure 5b is a photograph of a dissected pea flower revealing the stamens bearing the yellow pollens.

Legend of Figure 5b. Photograph of a dissected pea flower showing the stamens bearing yellow pollens

Mendel cultivated and tested some 29,000 pea plants between 1856 and 1863. He saw that there were variations in different characters of the pea plants: like height of the stem, tall or short; colour of the pea pods, green or yellow; morphology of the pea seeds, round or wrinkled; colour of the pea flowers, blue or white and so on Two different types of pea seeds, round and wrinkled are shown in Figure 6.

Legend of Figure 6. Photograph of wrinkled pea seeds (left) and of smooth round seeds (right). *The pea seeds are about 7mm in diameter.*

For performing crosses say between tall stem plants and short stem plants, Mendel removed the stamens, the pollen bearing male organs of the recipient flowers of the tall stem plants, leaving their ovaries intact. He then took out the stamens from the donor flowers of the short stem plants and deposited their pollens on to the ovaries of the recipient flowers. Surprisingly, when he collected the seeds that formed after crossing and allowed plants to be generated from these seeds, all the progeny plants had tall stems, no plant had short stem or stem of intermediate size.

Did it mean that in case of pea plants, the pollens only allowed fruiting, but did not contribute anything to the character of the resultant plant? That would be very different from what happened with date palms. Mendel was not convinced. He allowed the flowers of the resultant plants to be self-pollinated (allowed to get pollinated by pollens from their own stamen). When he sowed the seeds that were formed, about one-fourth of the plants that grew were short stemmed. Mendel counted 788 tall stemmed plants and 277 short stemmed plants. A representation of this experiment is given in Figure 7.

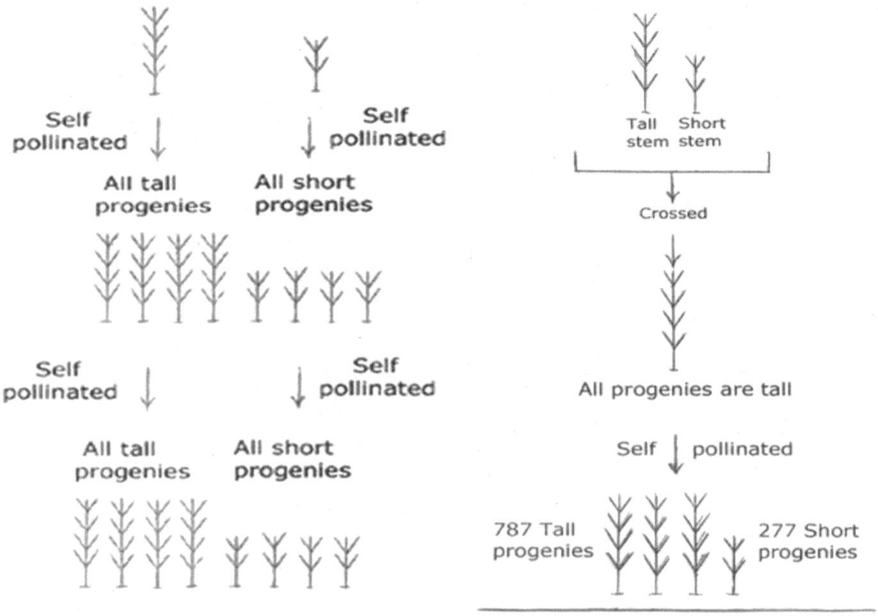

Legend of Figure 7. Left frame: Self pollination of tall plants or short plants. *Self pollination of tall plants always produces progeny plants that are tall. Similarly, self pollination of short plants always produces progeny plants that are short.* **Right frame: Cross between plants with tall stems and short stems.** *When the pollens from the flowers of a tall plant are applied to the destamenized flower of a short plant or the pollens from the flowers of a short plant are applied to the destamenized flower of a tall plant, all the progeny plants are tall.*

When these progeny tall plants (F1) are allowed to flower and self-pollinate, the next generation of progeny plants are a mixture of tall and short plants and they are in the ratio of 3:1 respectively. This figure is a composite reproduction of figures 1.1 and 1.2 from Chapter 1 of the book, Gene and its Engineering (ISBN:978-81-265-4928-3). Chapter 1 was contributed by H. K. Das. The book was edited by H. K. Das and published by Wiley India Private Limited in 2014.

Gregor Mendel

So, the hereditary information from the short plants were not really lost, but was only dormant in the first generation. Mendel obtained similar results when he performed new crosses and looked at the other characters of the pea plants like seed morphology (smooth or wrinkled) or colour of the pea pods or of the flowers. Mendel proposed that each character of pea was controlled by a hereditary "factor" which existed in multiple forms, but he had no idea as to what comprised a hereditary "factor" or how it controlled a particular character of a pea plant.

Mendel read his paper, "*Versuche über Pflanzenhybriden*" ("Experiments on Plant Hybridization"), at two meetings of the Natural History Society of Brno in Moravia, but the attending scientists failed to appreciate the importance of his work. Dejected Mendel reportedly told a friend, ""My time will come". It never came during his life time. It was cited just three times over the next thirty-five years. It was only in the early twentieth century, that Mendel's contribution was appreciated and he was termed as the father of modern Genetics.

In 1868 he was elevated to the post of abbot and his increased administrative responsibilities prevented him from continuing with experiments.

Mendel died in 1884, at the age of 61 from chronic nephritis.

3

JOHANNSEN'S GENE

We again wait, but this time the wait is for only two decades after Mendel's death. We move from Moravia in the Czech Republic to Copenhagen in Denmark. A professor is performing crossing experiments similar to that of Mendel, but with bean plants. This is Wilhelm Ludwig Johannsen, Professor of Botany and Plant Physiology in the University of Copenhagen.

Wilhelm was born in Copenhagen in 1857 in a poor family, which could not afford to give him formal higher education. He started serving as an assistant to a pharmacist when he was 15 years old. Seven years later he passed the pharmacist's examination and became a qualified pharmacist. He, however, did not enjoy the routine work of a pharmacist. In 1881, he left the pharmacist's job and joined as an assistant in the Carlsberg Laboratory in Copenhagen under the world renowned Chemist, Johan Kjeldahl. Dr. Kjeldahl appreciated Johannsen's keenness to study physiology of plants and gave him a free hand.

Johansen's work was recognized by the then plant physiologists and in 1892 he was appointed Lecturer at the Royal Veterinary and Agricultural University at Copenhagen. In a few years he became a Professor there. In 1905, he was invited to be a Professor of Botany and Plant Physiology in the University of Copenhagen. There Johannsen shifted his interest from plant physiology to plant heredity. His crossing experiments with bean plants yielded results very similar to that of Mendel.

By that time various researchers in the field of heredity had proposed different but nearly synonymous terms to describe the hypothetical hereditary

---

Foot note: Annexure to Chapter 3 may be seen to know about the terms genotype and phenotype that Johansen had also coined.

units. These were "essence" (Chalres Naudin, 1860), "factors" (Gregor Mendel, 1865), "gemmules" (Charles Darwin, 1868) and "pangene" (Hugo de Vries, 1889).

In 1909, Johannsen coined the term "gene" for the hereditary unit and that term found acceptance everywhere. Even today the universally accepted name of the hereditary unit is gene.

Wilhelm Ludvig Johannsen

Gene was still an abstract term then, with no clue as to what its nature really was and how it functioned.

In 1917, Johannsen was appointed the Vice-chancellor at the University of Copenhagen. He was a corresponding member of the Academy of Natural Sciences of Philadelphia.

Johannsen passed away in 1927.

Annexure to Chapter 3

Johannsen also coined the terms genotype and phenotype, genotype referring to the hereditary code inherited from the parents, while phenotype referred to the observed appearance. The coining of these two separate terms had a very important implication. Even if the genetic make up of an offspring is strictly governed by the genotype of the parents, the phenotype (external appearance) of the offspring may not be exactly similar to that of the parents.

The second important implication is that some genes may remain dormant in a particular generation, that is why the offspring is not an exact replica of either of the parents or mixture of the parents. Some of the characters might reflect characters of grand parents or great grandparents. On the other hand, some of the genes inherited from either parents might express (show up) in subsequent generations.

Johannsen later investigated why seeds of a particular species of plant would not germinate in all seasons, but would remain dormant. He discovered that dormancy of seeds could be broken by anesthetics like chloroform or diethyl ether.

4

GARROD'S STUDY OF ALKAPTONURIA, A DISEASE THAT APPEARED HEREDITARY

Another act of our story was being enacted across the English channel just a decade after the death of Mendel, even before Johansen coined the term gene. A physician was attending to patients in his hospital in London.

The physician was Archibald Edward Garrod. He was born in 1857 in London. His father was also a physician, but wanted his son to study business. Garrod, however, became interested in natural diversity, particularly of butterflies as early as 12 years of age and noticed the comparative rarity of female butterflies. He was sent to Marlborough College to study liberal arts, but did very poorly in both Latin literature and grammar. Recognizing his talent and inclination, his teachers encouraged him to go into the field of science and medicine.

Garrod studied at Christ Church in Oxford University and graduated with a First-class Honours degree in natural science in 1878. In 1880 he joined the St. Bartholomew's Hospital in London and graduated in medicine in 1884. Over the next several years he served in different hospitals in London and was later appointed the Regius Professor of Medicine at Oxford

In 1897 a mother came to Garrod's hospital with an infant, whose diaper was stained brownish-black. The ailment was known as alkaptonuria and was thought to be the result of some kind of urinary infection in infants. Three years later the same mother became pregnant again. When the new baby was born, Garrod noted the appearance of the brownish-black urine few hours after its birth. He suspected that the ailment was not the result of any infection, but was the manifestation of a hereditary defect.

The mother of the baby had married her first cousin. Neither she nor her husband had this ailment. Her grandfather had two sons, one son was her father, while the other son was her husband's father. The grandfather had the ailment alkaptonuria.

Garrod started looking for other patients with the same disorder and recorded the family history of these babies. He noticed that alkaptonuria always occurred in children of first cousins, whose one common grandparent had the same ailment. Garrod was convinced that first cousin marriage *per se* was not the cause, as he had seen that urine of children of first cousin couples, none of whose grandparents or great grandparents had this ailment, never changed its color on standing in air.

Interestingly, he also found that if the father's father of a woman had this ailment and the woman had married the son of her mother's brother, her baby's urine did not turn brownish-black. These observations converted Garrod's suspicion into conviction, Alkaptonuria in a child is the manifestation of a hereditary defect, which must be there in both parents, even if in a silent or symptomless form. .

In 1902 Garrod published his paper on alkaptonuria, where his vision went one step ahead of Mendel. Garrod envisioned that it was the chemical reactions that were going on in an individual human's body that differentiated one person from another, implying that Mendel's hereditary "factors" finally worked by controlling chemical reactions in an individual, whether it was a pea plant or a human being.

The title he gave to his paper was "The Incidence of Alkaptonuria: A Study of Chemical Individuality". Garrod wrote, "Owing, as I believe, to their chemical individuality, different human beings differ widely in their liability to individual maladies, and to some extent in the signs and symptoms which they exhibit".

Though Garrod guessed how the hereditary "factor" worked, its nature was still a mystery. Also a mystery was how the hereditary "factors" controlled chemical reactions in a living being.

Foot note: Annexure to Chapter 4 may be seen for analysis of the brwonish black urine of people afflicted with Alkaptouria.

Garrod was a member of the Royal College of Physicians, London, was elected a Fellow of the Royal Society and was made an honorary member of the American Association of Physicians, and of the Arztlicher Verein in Munich. Garrod served the army during World War I and was knighted in 1918.

He had varied interests and wrote a monograph entitled "A Handbook of Classical Architecture". He died after a brief illness in 1936.

Annexure to Chapter 4

Chemical analysis revealed later that the urine of a person afflicted with Alkaptonuria contains a chemical compound called homogentisic acid. Homogentisic acid is not present in the urine of a healthy person, because it is converted in the body of a healthy person into another chemical compound called 4-maleylacetoacetic acid through a chemical reaction controlled by a hereditary "factor". In a person afflicted with Alkaptonuria, the controlling hereditary "factor" is defective, hence homogentisic acid is not converted in the body into 4-maleylacetoacetic acid, but accumulates in the body and is excreted in the urine. When homogentisic acid is reacted upon by oxygen in air, a brownish-black chemical compound named alkapton is produced and the colour of the urine is changed to brownish black.

5

BERZELIUS'S CATALYSTS

Before we move forward any farther, we pause and look back for clues to understand how chemical reactions can take place in living beings, whose body temperature is the ambient temperature, or 37°C for warm blooded animals and humans. The problem is that many chemical reactions need high temperature to proceed. We, however, find that as early as 1835 a Swedish gentleman named Jöns Jacob Berzelius discovered the role of catalysts in chemical reactions. He found that chemical reactions that normally would need high temperature, can take place at a lower temperature, if a catalyst is present. Berzelius guessed that chemical reactions that go on in a living object might also be mediated by catalysts.

Berzelius was born in 1779 in a parish in Östergötland in Sweden. His father used to teach in a school, but died within 4 years after his birth. His mother also died when he was only 9 years old. Berzelius then went to live with his uncle whose wife was an alcoholic, so he had to spend most of his time away from home.

At the age of 13 Berzelius took admission in a Cathedral School. He tutored wealthy students to support him and pay his boarding fees. He, however, did not enjoy learning about religion and the church, but most of the time roamed around collecting insects. He wrote in his diary, "when I came home, I rearranged my insect collection which is crowding the room and now includes 369 specimens, many of them rare."

Berzelius did not do well in the school, but the Bishop recognized his talent and interest in nature and recommended him to be admitted in Uppsala University. Berzelius chose to study medicine, but could not continue for want of money. In 1798 he won a three–year scholarship and resumed studying medicine.

Jöns Jacob Berzelius

Chemistry was an integral part of medicine in Uppsala and Berzelius was fascinated by the work of great chemists and tried to repeat their experiments. As he was successful in repeating the experiment of Carl Scheele, the discoverer of oxygen, he wrote in his diary, "I have seldom experienced a moment of such pure inner delight as when the glowing stick held in the gas [oxygen] ignited and illuminated my windowless laboratory with unusual brightness." When Alessandro Volta constructed the first electric battery in 1800, Berzelius repeated the construction with great satisfaction.

As his three-year scholarship ended in 1801, he financed his studies by working as an apprentice in a pharmacy and subsequently as an assistant of a physician. He analysed mineral water for a mine owner in exchange for meals. He obtained the M.D. degree in 1802 and started practicing medicine. In 1807 he left his practice and joined as a professor in Chemistry and Pharmacy at the Karolinska Institute. There he wrote about his concept of catalysts "The agent causing the conversion of substances does not participate in the new compounds formed but remains unchanged, thus operating by means of an internal power, the nature of which is still unknown".

In other words, a catalyst is a substance that "mediates a chemical reaction or enhances the rate of a chemical reaction without itself undergoing any permanent chemical change". An example is the catalytic converter in a car. The converter contains platinum, which acts as a catalyst to convert the toxic carbon monoxide in the exhaust gas into carbon dioxide. In the absence of the catalyst, the conversion would have needed a much higher temperature.

Subsequently, Berzelius wrote about involvement of catalysts in chemical reactions in living objects, (the English translation has been done by Joseph S. Fruton and Sofia Simmonds) "We have justifiable reason to suppose that, in living plants and animals, thousands of catalytic processes take place between the tissues and fluids and result in the formation of the great number of dissimilar chemical compounds, for whose formation out of the common raw material, plant juice or blood, no probable cause could be assigned. The cause will perhaps in the future be discovered in the catalytic power of the organic tissues of which the organs of the living body consist."

Berzelius's guess about catalysis in living objects was absolutely correct, we know today that there are many chemical reactions that occur in the human body or plants which are mediated by catalysts. For example, when we inhale air, the oxygen in it gets converted into carbon dioxide in the body and we exhale the carbon dioxide. This carbon dioxide is the result of reactions in our body at 37°C (our body temperature) between oxygen and substances like sugar or starch (in wheat flour or rice) or other food that we eat. This reaction with oxygen is possible in our body at 37°C, because of the intervention by catalysts. In contrast, oxygen in air does not react with sugar or rice powder or wheat flour kept in a jar and produce carbon dioxide even in summer, when in many parts of India the day temperature rises to 37°C and higher. Besides reactions with oxygen, all kinds of other reactions go on in the body mediated by catalysts.

Mothers milk or milk formulations, the only source of nutrition of a new born child, is converted into all the diverse parts of the child, namely, skin, hair, muscle, nerves, bones, the different organs etc. All these are because of myriads of chemical reactions in a body, now known as biochemical reactions, that take place at the body temperature mediated by catalysts, which may be termed as biocatalysts.

The first practical use of biocatalysts occurred more than 5,000 years ago, when people made cheese by storing milk in stomach taken out of dead animal. The biocatalyst present in the inner layer of animal stomach has later been identified and named as "rennet".

Around the late 17th and early 18th centuries, people knew that the secretions from the stomach could digest meat, while saliva could convert starch into sugars. Of course, it was not understood then that biocatalysts were involved in these processes.

Foot note: A healthy person's urine is not brownish black, because of intervention of a biocatalyst. See Annexure to Chapter 5.

In 1840 Berzelius was made the Knight of the Order of Leopold. Berzeliusskolan, a school situated next to his alma mater, Katedralskolan, has been named after him.

Berzelius died in 1848. His portrait appeared on a series of postage stamps in 1939, commemorating the bicentenary of the founding of the Swedish Academy of Sciences.

Annexure to Chapter 5

Chemical reactions can sometimes take place spontaneously also without intervention of a catalyst. An example of a spontaneous chemical reaction is the interaction of homogentisic acid (colourless), present in the urine of alkaptonuria patients, with oxygen in air, to produce the brownish-black substance alkapton.

Homogentisic acid is not present in the urine of a healthy person, as it is converted in the body into 4-maleylacetoacetic acid. The conversion in the body also involves oxygen, but interestingly the product is different. A biocatalyst intervenes and a different reaction takes place leading to a different product.

6

PASTEUR'S VITAL FORCE

Almost three decades went by after the discovery of catalysis, we have travelled from Sweden to France, the country famous for its wines. The owner of a winery has requested the great French scientist, Louis Pasteur to find out as to why some times the wine would become sour. Pasteur looked at the different wine samples under the microscope and noticed the presence of living organisms, which were not visible with the naked eye. These microscopic organisms were named yeast. The wines that had gone sour contained microscopically small organisms that looked very different from the ones that produced good wine (see Figure 8). If the juice was boiled at 100°C immediately after crushing the grapes, no wine was produced. Pasteur inferred that the microscopically small organisms present in grape juice were responsible for converting the grape juice into wine, as these must have been killed by boiling.

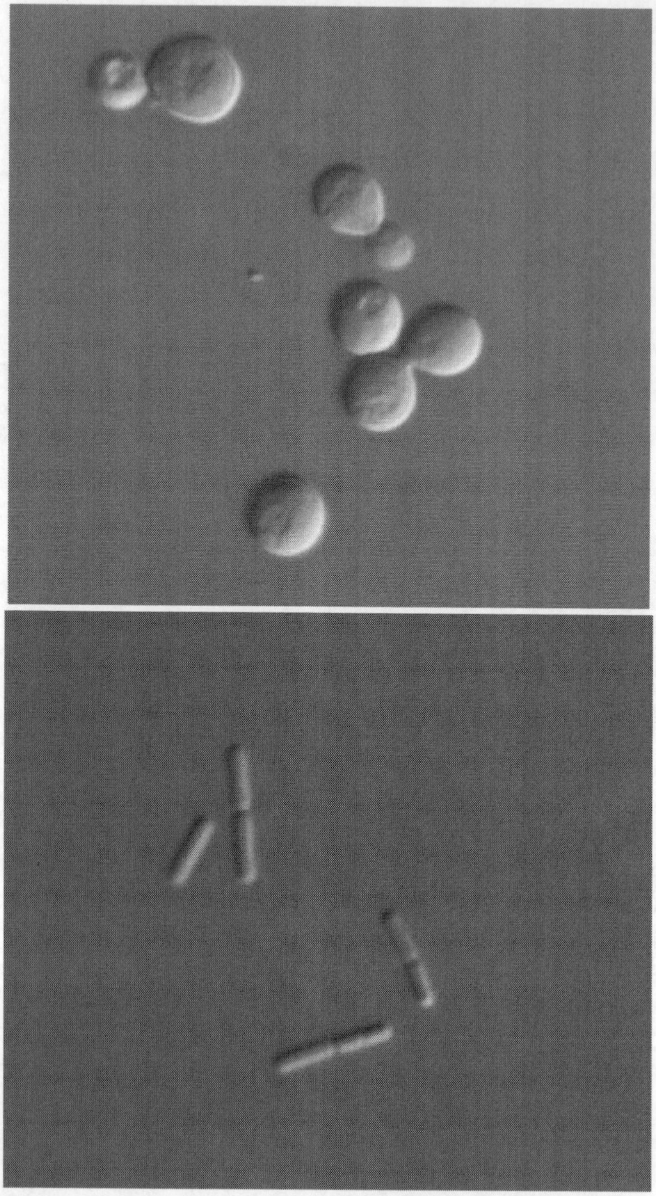

Legend of Figure 8. Photographs taken through microscope. *Magnification, 1000 fold. Top photograph: Yeast that produce good quality wine from grape juice. The bottom most image in the top photograph shows a single yeast. These yeasts multiply by "budding", the three different stages of multiplication by budding captured in the photograph can be seen. Bottom photograph: microorganisms that produce sour wine. One single microorganism can be seen in left most position. These microorganisms multiply by elongation and division. Three such division events can be seen in the photograph.. The photograph was taken by Preeti Dabas in the laboratory of Professor Nimisha Sharma of the School of Biotechnology, Guru Gobind Singh Indraprastha University, New Delhi, India.*

Louis Pasteur was born in Dole, France in 1822. His father, after retiring from Napoleon's army, was earning a living by tanning leather, a business initiated by his great grandfather in 1763. Louis Pasteur's father did not have any formal education, but earnestly desired and strived to give his son the best possible education. The boy Louis was an excellent artist, but he had to stop painting at the age of 19, as his father did not approve of his son pursuing this profession. Instead, he wanted Louis to have education in a University and become a Professor at the local College in Arbois. Louis, however, was a very ordinary student in school, though he excelled in drawing and painting. He failed in the entrance examination of the Ecole Normale Superieure in Paris, the well known seat of higher education, but did not give up. He succeeded in his second attempt and earned both his Master's and Doctoral degrees in Chemistry from there. In 1848 the University of Strasbourg appointed Louis Pasteur as Professor of Chemistry and in 1854 University of Lille appointed him as the Dean of Science. Pasteur's next assignment was in 1857 as Director in charge of Science studies at the Ecole Normale Superieure in Paris, the same Institute where he failed to get admission as a student in his first attempt.

Louis Pasteur

In 1864, Pasteur discovered that heating the grape juice to 55°C would kill the microscopically small organisms that produced the sour tasting material, but the microscopically small organisms that produced good wine would be unaffected. "Chance favours invention only for minds prepared for discoveries by patient study and persevering efforts", Pasteur had said.

Pasteur observed that these microorganisms multiply rapidly in grape juice. He purified the two kinds of microorganisms and allowed these to multiply separately. He observed that the progenies of the microorganisms that were producing good quality wine were always producing good quality wine and were not affected on heating at 55°C, while the progenies of the microorganisms that were producing sour wine were always producing sour wine and were always killed on heating at 55°C. Thus the laws of hereditary transmission of genetic characteristics worked with microorganisms also.

Pasteur was a great scientist, a biologist and a chemist combined, but he believed that production of alcohol and formation of wine were mediated by a "vital force" present in the yeast cells and that it was an "act correlated with the life and organization of the yeast cells". Pasteur could not conceive that catalysts, as predicted by Berzelius, could be involved in the production of alcohol and formation of wine caused by microscopic living organisms like yeast.

While at Ecole Normale Superieure, Pasteur shifted his studies from wine to diseases and discovered that many of the diseases were also caused by microorganisms, which he called "germs". His "germ theory" has since then become central to our understanding of disease. Pasteur also discovered "vaccines" as he found that the cholera causing germs which had been weakened and made non-infectious by long exposure to air, when injected into chicken, made them immune to infectious cholera germs.

At the request of the Ministry of Agriculture in France, Pasteur started investigating silkworm diseases and found that most of these diseases were also caused by germs. In 1867 he joined as Professor of Chemistry in Sorbonne, the world famous University of Paris, but continued his work on germs and infectious diseases. He suffered a stroke in 1868 that caused partial paralysis, but the work he was doing on vaccines against cholera, small pox, rabies and anthrax continued. A research institute was established in Paris in 1887 dedicated to the study of biology, micro-organisms, diseases, and vaccines and was named Institut Pasteur, of which Louis Pasteur was the first Director. Louis Pasteur was awarded the Cross of the Legion of Honor, Rumford Medal of the Royal Society of England, the Grand Prix medal from the Exposition Universalle and the Grand Cordon of the Legion of Honor. Pasteur died in1895, due to several strokes.

7

KÜHNE'S ENZYME

It was 1875, Pasteur was still alive, from France, we go across the border to Germany and find that the physiologist, Wilhelm Kuhne has opined against Pasteur's view that a "vital force" present in the yeast cells was responsible for alcohol production and has voiced in favour of Berzelius's idea of involvement of biocatalysts in chemical reactions in living objects. Kuhne coined the term "enzymes" for the biocatalysts carrying out alcohol production by yeast. The term enzyme was derived from a Greek word meaning "leavened".

In 1876 Kuhne discovered the enzyme trypsin from the secretions of the pancreas and demonstrated its ability to digest proteins.

Wilhelm Kühne

Wilhelm Kuhne was born in 1837, the son of a wealthy German merchant of Hamburg. In 1856 he obtained his Ph.D. degree in Physiology from the University of Göttingen. He continued his education in Jena, Berlin and Vienna and finally joined as a faculty in Rudolf Virchow's Berlin Institute in 1861. In 1868 he was appointed the Professor of Physiology at Amsterdam University and in 1871 at Heidelberg University. He was elected member of the Royal Swedish Academy of Sciences and was awarded the Nobel prize.. Wilhelm Kühne died in 1900.

8

BUCHNER'S ALCOHOL PRODUCTION WITHOUT LIVE YEASTS

Wilhelm Kuhne believed that biocatalysts, rather than a "vital force" present in the yeast cells, converted the sugar present in grape juice into alcohol, but he did not have any concrete proof. Two decades later Eduard Alois Buchner, a chemist from his own country, observed that conversion of sugar to alcohol could be achieved by adding extract from crushed yeasts that contained no living yeast in it. This was the conclusive nullification of Pasteur's notion that the conversion "was an act correlated with the life and organization of the yeast cells".

Buchner inferred that indeed a biocatalyst present in the extract from crushed yeast was doing the job and named it "zymase". So "zymase" was an "enzyme". It was revealed later that "zymase" actually comprised a number of different "enzymes". Henceforth all biocatalysts, whether from animal or plants or microorganisms, were classified as "enzymes", but nobody knew what the chemical nature of an enzyme was, or what was its relationship with the gene.

Eduard Buchner

Eduard Alois Buchner was born in 1860 in Munich, Germany in an educated family. His father was a Professor of Forensic Medicine and Obstetrics and also the editor of the journal *Ärztliches Intelligenzblatt* . His father, however, died when he was only twelve years old. His brother, Hans, who was ten years older, supported him for his education. Buchner graduated in Munich and then studied Chemistry. But, he had to discontinue his studies for want of money and had to work in canneries for four years. His elder brother Hans Buchner, who was by then a Physician and Bacteriologist, helped him out again and in 1884 he resumed studying Chemistry. He also worked as a Laboratory Assistant of the famous chemist, Adolf von Baeyer.

Buchner obtained his Ph.D. degree in 1888 and was promoted to the post of Teaching Assistant in 1890. Buchner was appointed a Lecturer of Chemistry in the University of Kiel in 1893. In 1896 at the age of 36 he was appointed Professor of Analytical and Pharmaceutical Chemistry in the University of Tubingen.

His brother Hans Buchner was in Munich then and Eduard would always spend the holidays with him. There Eduard would prepare extract of yeast for Hans, who would use it for his experiments. Edward would grind yeast with quartz sand in a mortar and would squeeze the resultant paste through canvas and save the extract that came out for use by Hans. Eduard remembered his days in the canneries, when he would add a thick syrup of sugar to marmalade to prevent it from rotting before canning. He did the same thing to prevent the yeast extract from rotting and stored it for subsequent use by Hans. The next day he observed bubbles in the yeast extract and froth on its surface. It smelt of alcohol. Eduard examined the yeast extract carefully under the microscope, but could not detect any live yeast cells. So conversion of sugar into alcohol really did not need live yeast.

On January 9, 1897 Eduard Buchner sent his paper, *Uber alkoholische Garung ohne Hefezellen* (On alcohol production without yeast cells), to the editors of the reputed scientific journal *Berichte der Deutschen Chemischen Gessellschaft*. His paper was published on March 3, 1897. By 1902 Eduard had published altogether 17 papers describing in great detail different aspects of alcohol production using extract of yeast cells, but without any live yeast. His work established a new branch of Science named Biochemistry.

Eduard Buchner was awarded the Nobel prize for his work. He said in his Nobel lecture, "The active agent in the expressed yeast extract appears rather to be a chemical substance, an enzyme, which I have called 'zymase'. From now on one can experiment with this just as with other chemicals." He continued, "we are seeing the cells of plants and animals more and more clearly as chemical factories, where the various products are manufactured in separate workshops. The enzymes act as the overseers".

In 1914 the World War 1 erupted and Eduard Buchner was drafted in the German army. Two years later he was relieved from the army and he devoted himself to his laboratory work and hobbies of hunting and mountain climbing. Eduard Buchner, however, volunteered again to serve in the German army in 1917, when the United States of America stepped into the war in favour of the Allies. Unfortunately, Eduard was hit by a shell fragment in the Romanian front and died at the age of 57 on August 13, 1917. A tragic end of a Nobel prize winner.

9

SUMNER'S DISCOVERY - ENZYMES ARE PROTEINS

Eduard Buchner's paper on alcohol production without yeast cells was published in 1897. By around 1900 it was established that myriads of chemical reactions that go on in a living object, whether it was a plant or human or a microorganism, were possible because of no mysterious "vital force", but because of the involvement of biocatalysts, which have been named enzymes. Though it was guessed that the enzymes were chemical entities, nobody had an inkling as to what was the nature of the enzymes or what the enzymes were composed of.

We wait for about two decades and travel across the Atlantic to the U.S.A. We come across one James Batcheller Sumner, who was then an Assistant Professor of Biochemistry at Cornell Medical School in Ithaca, New York state, U.S.A. He had read about Buchner's work on the enzymes and had the conviction that there was nothing mysterious about the enzymes and longed to embark on the task of unravelling the nature of enzymes. His teaching load, however, was very heavy and by then he had taught for three years without any spare time for indulging in any kind of research. He also did not have any body to assist him in the laboratory. In addition, he had a physical handicap. Sumner was fascinated by fire arms and often went hunting with his friends. It was in one of those hunting trips, Sumner was accidentally shot by his friend in the left arm. and his arm had to be amputated just below the elbow. Sumner, however, had indomitable spirit and extraordinary courage. He was originally left handed, but he learned to do everything with his right hand and was firm in his determination to achieve his objective.

James Batcheller Sumner was born in 1887 in Canton, Massachusetts, U.S.A. He was a direct descendent of the anti-slavery campaigner, Charles Sumner. He was the son of a wealthy owner of a large country estate and the grandson of a owner of a textile factory. As was the norm those days, he was sent to a Latin school, but he exhibited no interest in learning classical languages. Instead, he enjoyed the Physics and Chemistry classes.

James B. Sumner

Sumner obtained his Bachelor's degree from Harvard University in 1910. He then worked in the cotton knitting factory of his uncle for some time, but did not like it very much. He taught in a college for a while, but did not enjoy it either . In 1912, he went to study Biochemistry in Harvard Medical School. Professor Otto Folin was impressed with his interview, but was apprehensive that a one handed man can not successfully pursue a career in Biochemistry. He advised Sumner to study Law. Sumner persisted and within two years obtained his Ph.D. degree in Biochemistry. In 1914, Sumner was offered the position of Assistant Professor of Biochemistry at Cornell Medical School in Ithaca, New York state, U.S.A.

Early on, Sumner reasoned that if he wanted to characterize the chemical nature of enzyme, he must be able to obtain the particular enzyme in a very pure state. For that he should chose an enzyme whose source material was abundantly available. The enzyme urease in jack bean meal was his choice. Urease converts urea, a substance that occurs in our urine, to ammonia and carbon dioxide in presence of water. If urea is added to a paste of jack bean meal, one can smell of ammonia, the familiar odour of smelling salt. Crystals of chemical substances are considered to be of good purity. A common example would be crystals of cane sugar (Figure 9) isolated from sugar cane juice, which we consume every day. Sumner set his goal to crystallize the enzyme urease. It must be understood, that till then nobody had ever crystallized an enzyme, or nobody even thought that enzymes could be crystallized. Sumner was unsuccessful in his mission for several years and many of his colleagues believed that he was trying to achieve something that was impossible. Sumner did not give up, finally in 1926 he demonstrated that the enzyme urease could indeed be crystallized. Crystals of urease can be seen in Figure 10.

Once crystallization was successful, Sumner went about finding out the chemical nature of the enzyme urease. Specific chemical tests revealed that urease was nothing but a protein. Biochemical reagents that degraded proteins, abolished the ability of urease to convert urea into ammonia and carbon dioxide, thus establishing beyond any reasonable doubt that the enzyme urease was indeed a protein. Other enzymes were subsequently crystallized by Sumner himself and other scientists by following Sumner's method. Those enzymes were also found to be proteins.

Sumner was awarded the Nobel prize in Chemistry in 1946 for his work on the enzyme urease. During his Nobel lecture, Sumner said, "I wish to tell next why I decided in 1917 to attempt to isolate an enzyme. At that time I had little time for research, not much apparatus, research money or assistance. I desired to accomplish something of real importance. In other words, I decided to take 'a long shot'. A number of persons advised me that my attempt to isolate an enzyme was foolish, but this advice made me feel all the more certain that, if successful, the quest would be worthwhile."

Sumner had earned the reputation of an excellent teacher. His philosophy of teaching was, as he had expressed once, "The most important things I have tried to give my students are a curiosity to discover the world around them and the integrity to look only for the truth. Students must learn what has gone before them in the past and gain broad rather than too specialized backgrounds if they are really to know where they are going and what they are doing." He served as Assistant Professor for fifteen long years, when he was promoted to the position of Professor. In 1947 Sumner was appointed as the Director of the newly established Laboratory of Enzyme Chemistry.

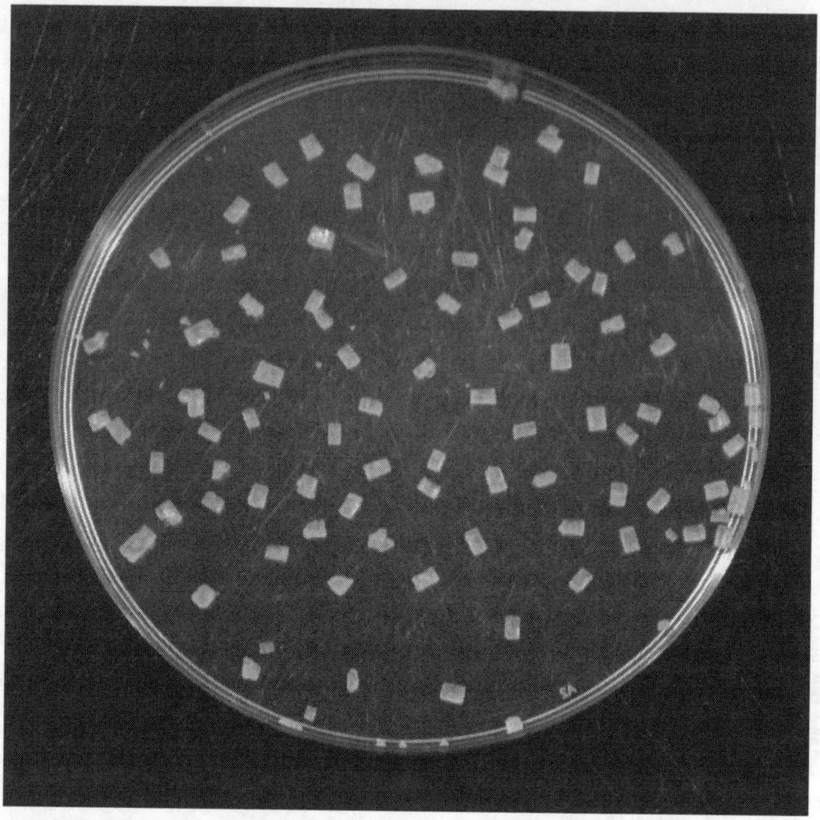

Legend of Figure 9. Crystals of cane sugar. *The actual width of the crystals is about 2mm.*

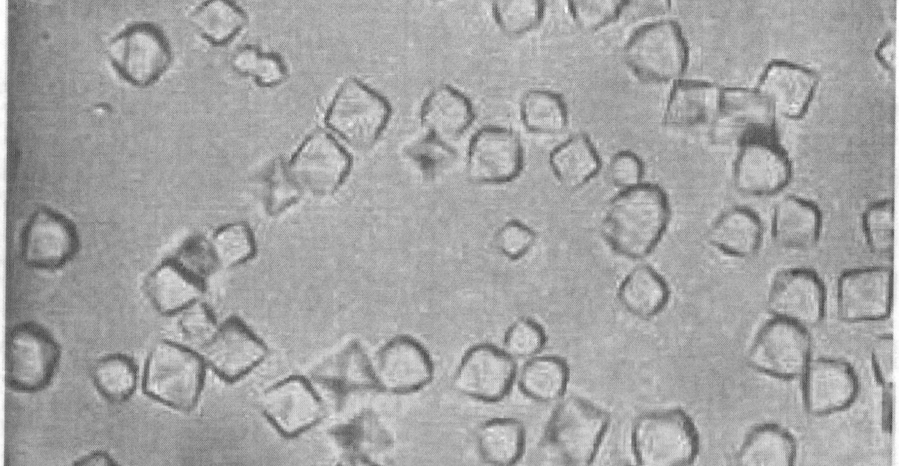

Legend of Figure 10. Crystals of the enzyme urease.

Sumner was a great sportsman too. His desire to excel in sports like tennis, skiing, skating, billiards and clay pigeon shooting was in no way deterred by the loss of one arm and he became a competitive amateur player. He once won the Championship of the Cornell Faculty Tennis Club. Many years later, when he went to receive the Nobel prize in Stockholm, King Gustav of Sweden asked him how he managed to throw the tennis ball up and also serve with just one hand. Sumner demonstrated to the king by actually doing it.

Sumner had many other interests. He enjoyed canoeing and mountain climbing. He was an excellent photographer, a very good cook and a versatile linguist, reading and speaking German, French, and Swedish, besides his mother tongue, English.

In March 1955, Sumner fell ill and was diagnosed of being afflicted with cancer. A symposium was arranged in May, 1955 to honour him. After dinner, Sumner delivered a short but witty speech, which charmed every body. He did not let any body realize that he was in excruciating pain. That was the last symposium he attended, as he could not recover and died in August that year.

10

MULDER DESCRIBED THE MAIN CONSTITUENT OF ALBUMIN TO BE "PROTEIN"

Sumner had established that enzymes are protein in nature. We also know that fish or meat or pulses contain proteins, but what is the nature of proteins? As we go back in time, we find that the term "protein" was first used in 1838 by one Jerardus Johannes Mulder in a publication in French, the English translation of the title being, "On the composition of some animal substances" . We travel from New York, U. S. A. to Utrecht, the century old religious centre in central Netherlands in Europe. Mulder was born there in 1802. His father was a surgeon and wanted his son to become a surgeon. Jerardus got his degree in medicine in 1825. Within two months he got another degree in Pharmacy.

He practiced medicine in Amsterdam and subsequently in Rotterdam. Simultaneously, he taught Botany in the Batavian Society for Experimental Philosophy. In 1828 Mulder became a Lecturer in Botany, Chemistry, Pharmacy and Mathematics in the Clinical School of Rotterdam, but continued his medical practice also. In 1835 he discontinued his medical practice and devoted full time in teaching, research, editing Chemistry journals and writing a Textbook for Chemistry students. Mulder was appointed Professor of Chemistry at the University of Utrecht in 1840.

Gerardus Johannes Mulder

Mulder analysed albumin from the white portion of hen egg and discovered that the main constituent was very much different from other naturally occurring substances like starch. He described the main constituent of albumin to be "protein". He, however, could not go any further to elucidate the nature of proteins.

Mulder was aptly regarded as the founder of modern Chemical Science in the Netherlands. He was elected a foreign member of the Royal Swedish Academy of Sciences in 1850.

Later in life, Mulder became active in politics and became one of the leaders of the conservative movement against the liberal administration of the then prime minister, which resulted in the fall of the government in 1853. Mulder, however, suffered a mental breakdown in 1854 and resigned from the Professor's post, but the King of Netherlands refused to accept his resignation. He went on leave, resuming his teaching and research activities after a year, as he recovered from his ailment. Around mid 1870s Mulder's health started deteriorating and subsequently he became blind. He could not work in the laboratory any more. Nevertheless, he was mentally very active and dictated some books on the state of higher education in Netherlands. He passed away in 1880.

Foot note: Annexure to Chapter 10 contains the elemental composition of starch and protein. The Annexure also mentions about the controversy as to whether the term protein was actually coined by Mulder or Berzelius.

Annexure to Chapter 10

While starch is composed of carbon, hydrogen and oxygen, albumin contained nitrogen and sulphur in addition to carbon, hydrogen and oxygen. We know today that proteins are present in all living beings, be it animal or plant or microorganism. Some Science Historians say that Jöns Jacob Berzelius, coined the term "protein" and Mulder made use of it. Mulder, however, had written, "Berzelius communicated with me concerning the published results and gave me good advice for which I express my sincere thanks". This implied that Berzelius communicated only after seeing the publication of Mulder, so there was no question of Mulder using a term in his publication that was suggested or coined by Berzelius. Mulder clarified that he derived the word protein from the latin word *primarius* meaning primary or chief or distinguished.

11

SUGGESTION OF HOFMEISTER AND FISCHER THAT PROTEINS CONSIST OF MANY AMINO ACIDS LINKED TO EACH OTHER

Jerardus Johannes Mulder discovered that proteins were different from starch as proteins contained nitrogen, sulphur, carbon, hydrogen and oxygen, while starch contained only carbon, hydrogen and oxygen. That did not tell much. It took us long six decades and travel from Netherlands to Germany to know more about the composition of proteins. It was 1902, a conference was being held in Karlsbad, Germany under the auspices of the Society of German Scientists and Doctors. Franz Hofmeister, the well known protein chemist, was invited to deliver the plenary lecture on structure of proteins. Hofmeister opined there that proteins were linear arrays of many different amino acids.

linked by chemical bonds. Hofmeister's inference on the composition of proteins was drawn from his earlier observation that proteins we eat, irrespective of its source, are converted in the stomach and the intestine first into peptones (smaller proteins with lesser number of amino acids) and subsequently into amino acids.

Franz Hofmeister was born in 1850 in the city of Würzburg in northern Bavaria in Germany. His father was a respected and prosperous Physician, who established his medical practice in Prague, which is now the capital of the Czech Republic. Franz also was destined to be a Physician, but while he was studying Medicine, he exhibited special aptitude in Chemistry. He completed study of Medicine in 1872. In 1879, he obtained his *Habilitation* in Physiological Chemistry, the qualification to teach in a University.

In 1896, he became Professor of Physiological Chemistry at Strassburg, where he remained till his retirement in1919.

He spent his retired life in his birth place Würzburg, where he served as Honorary Professor of Physiological Chemistry. Hofmeister was a modest and gracious person, brilliant in conversation. He was a happy and healthy family man, who enjoyed his leisure painting with water colours and listening to music.

On the same day in 1902, in the same Karlsbad meeting, another famous German chemist, Emil Fischer independently suggested that proteins are polypeptides, in which amino acids are linked to each other. The smallest peptide is two amino acids linked by one chemical bond (amino acid 1 – chemical bond – amino acid 2). Polypeptide means a compound containing a large number of amino acids connected by many chemical bonds (amino acid 1 – chemical bond – amino acid 2 – chemical bond amino acid 3 – chemical bond – amino acid 4 – so on).

Fischer hydrolysed casein (milk protein) by boiling with acid and obtained a mixture of various amino acids, proving that proteins are composed of amino acids. He methodically approached the problem by first isolating and identifying each of the 18 different amino acids that constitutes casein..

Hermann Emil Louis Fischer was born in 1852, at Euskirchen in Cologne district of Germany, which was then a part of the Prussian Empire. He was a good student and had passed the High School Examination with distinction in 1869.

Emil wanted to go to college and study the natural sciences, but his father Laurenz Fischer, who was an industrialist running breweries, dyestuff making units and lumber business, wanted his son to assist him in the family business. Emil, however, proved a failure in business. As Emil had written in his autobiography, Emil's father said, "he was too stupid to work in the family business, so he better go study science".

Emil obtained Ph.D. in Chemistry in 1874 from the University of Strasbourg. He then worked as a Research Assistant to the famous chemist Adolph von Baeyer and moved with him to Munich in 1875. In 1878 he became a Privaydozent (unpaid Lecturer) and in 1879 an Assistant Professor.

Emil Fischer

Emil Fischer moved to Erlangen in 1882 as Professor of Chemistry. In 1892 he became the Director of the Chemistry Institute in the University of Berlin and continued to be there till his death.. This Institute was then the largest and most prestigious Chemical Institute in Germany, which attracted students from all over Europe, North America and Japan.

During his vacation in Bavaria, Fischer used to go for walk early in the morning in the Black Forest and there he would often get entangled with spider's web. His curiosity about the chemistry of spider's web led him to discover that webs were protein in nature.

During the period 1884 to 1900 Fischer worked on the chemical nature and structure of carbohydrates and sugars, which gave him the Nobel prize in 1902. Fischer was a member of the Prussian Academy of Sciences and was awarded honorary doctorate degrees by the Universities of Christiania, of Cambridge (England), of Manchester and of Brussels. Fischer was elected a Foreign Member of the Royal Society in 1899.

During the World War I, Fischer was appointed as the scientific adviser to the German Government, when he tried to produce synthetic food for the army. The German Chemical Society instituted the Emil Fischer Memorial Medal in his honour. The Nobel Prize winner, Richard Willstätter, who was Emil Fischer's colleague, had described Fischer as an "unmatched classicist, master of organic-chemical investigation with regard to analysis and synthesis and as a personality, a princely man." Fischer succumbed to cancer in 1919.

Let us now see what amino acids look like. Most amino acids have the same linear structure with the same two ends, we would call these as N-end and C-end. The N-end has a group that is basic in nature and the C-end has a group that is acidic in nature. . The difference between one amino acid and the other is in the side chain attached to this same linear structure. A schematic representation of some of the amino acids present in proteins is shown in Figure 11.

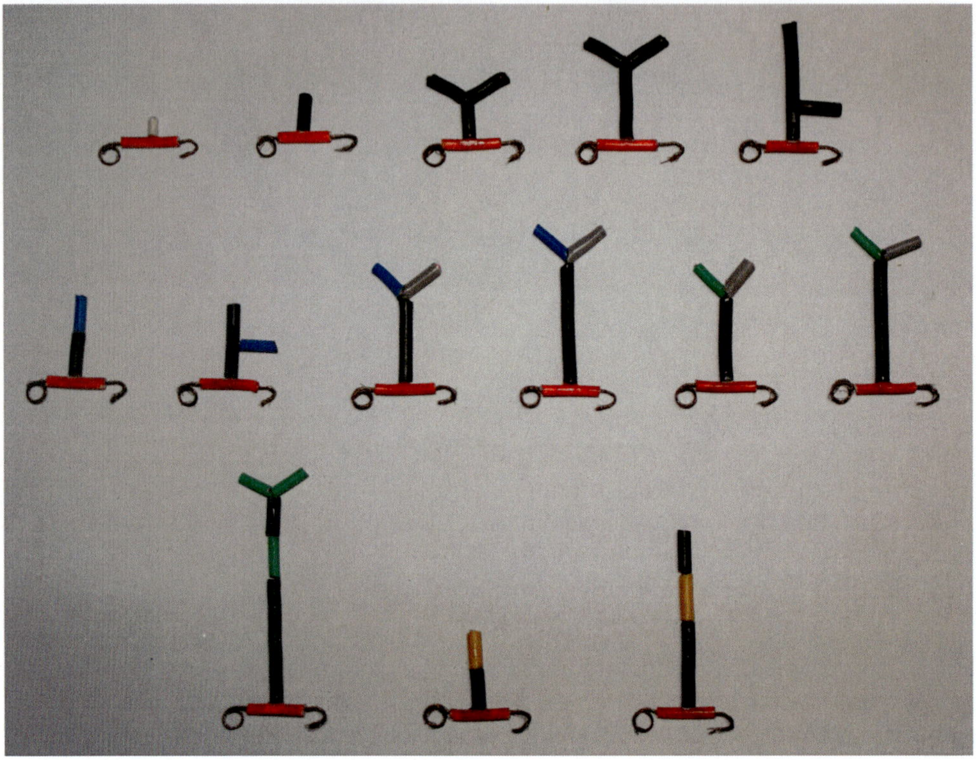

Legend of Figure 11. Schematic representation of some of the amino acids present in proteins. *The ring in the left end represents the basic group (also known as the N end), while the hook in the right end represents the acidic group (also known as the C end) . The red portion flanked between the ring and the hook, is a common feature in each amino acid, which contains a combination of one carbon atom and one hydrogen atom. Top row from left to right: Glycine, the small white butt in the middle of the red portion represents an additional hydrogen atom; Alanine, the side chain shown in black that comprises a carbon atom and three hydrogen atoms has replaced the hydrogen atom (white butt) in glycine ; Valine, the side chain is a branched one that contains three carbon atoms and seven hydrogen atoms; Leucine: the side chain is branched with a longer stem, contains four carbon atoms and nine hydrogen atoms; Isoleucine: side chain has the same four carbon atoms and nine hydrogen atoms, but the branch configuration is different; Middle row from left to right: Serine: the side chain is similar to alanine, but has an additional oxygen atom shown in blue; Threonine: the side chain is similar to that of serine, but is branched and has an additional carbon atom and two additional hydrogen atoms; Aspartic acid: the side chain is similar to leucine, but the terminal V structure comprising three carbon atoms and seven hydrogen atoms is replaced by a second*

carboxyl group comprising one carbon atom, two oxygen atoms and one hydrogen atom, which is shown by the blue and grey V structure; Glutamic acid: also has a second carboxyl group similar to Aspartic acid in the side chain, but the stem of the side chain is longer having an extra carbon atom and two hydrogen atoms; Asparagine: the second carboxyl group of aspartic acid is replaced by an amide group shown as green and grey V structure, comprising one carbon atom, one oxygen atom, one nitrogen atom and two hydrogen atoms; Glutamine: the second carboxyl group of Glutamic acid is replaced by an amide group; Bottom row from left to right: Arginine: contains three amino groups (shown green) in the side chain; Cysteine: similar to serine but the oxygen atom in the side chain is replaced by a sulphur atom shown in orange colour; Methionine: also has sulphur atom in the side chain, but with more carbon atoms and hydrogen atoms.

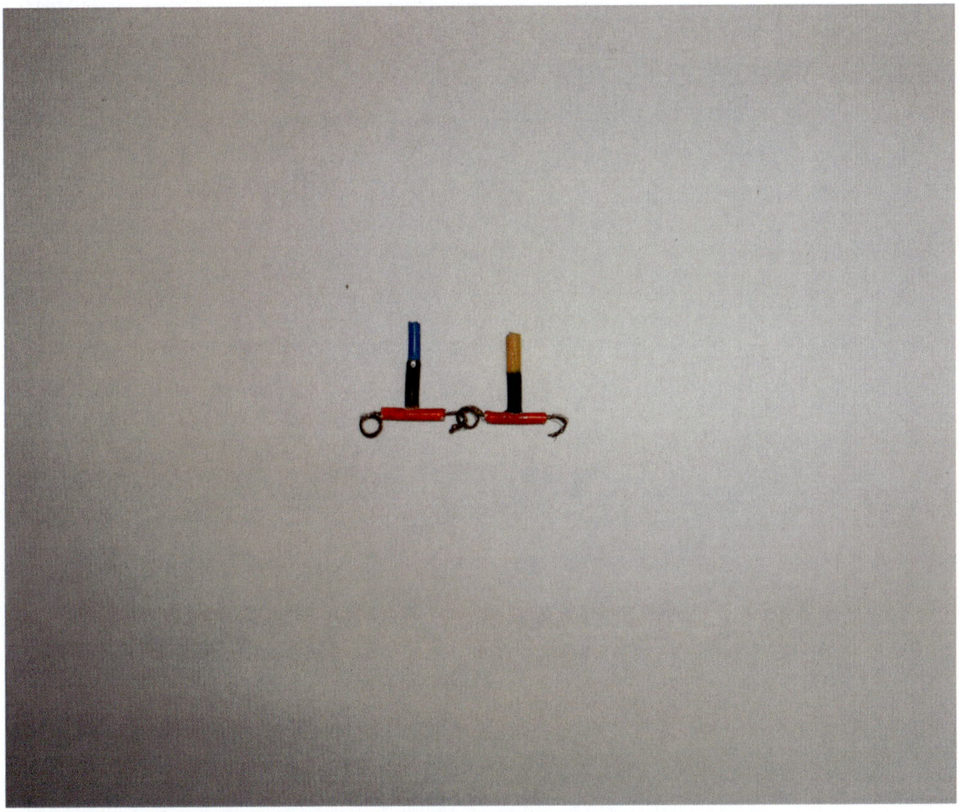

Legend of Figure 12. Schematic representation of a chemical bond between the amino acids Serine and Cysteine.

Foot note: Annexure to Chapter 11 contains the elemental composition of the N-end and C-end of amino acids. The Annexure also mentions the chemistry of the linking of one amino acid with the other during protein synthesis and the success story of the chemical synthesis by Emil Fischer of a protein containing eighteen amino acids.

The acidic C-end (shown as hook in Figure 11) of one amino acid reacts with the basic N-end (shown as ring in Figure 11) of the other amino acid forming a chemical bond between the two amino acids, resulting in a chain of two amino acids, which is called a dipeptide.

A schematic representation of a dipeptide connected by a chemical bond between the amino acids Serine and Cysteine is shown in Figure 12

Next the N-end of the amino acid 3 reacts with the C-end of the amino acid 2, thus forming a chain of three amino acids. Similarly, the N-end of the amino acid 4 bonds with the C-end of the amino acid 3 resulting a chain of four amino acids. Thus grows the chain finally producing a protein comprising a long chain of many amino acids, which in the language of Emil Fischer, a protein is nothing but a polypeptide. Figure 13 shows the schematic representation of a polypeptide chain of many amino acids.

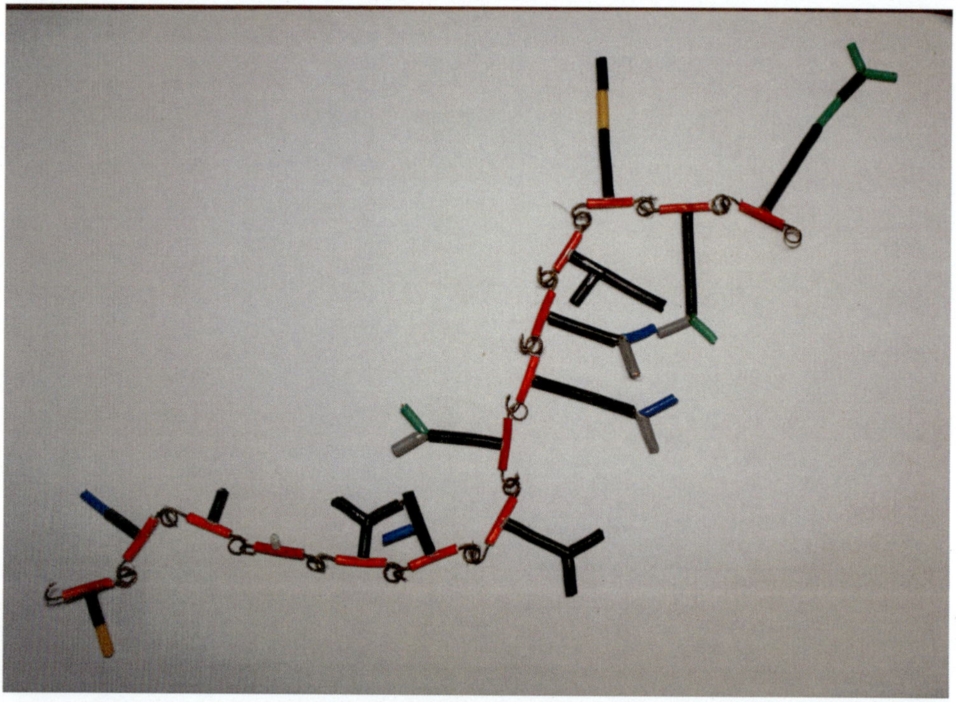

Legend of Figure 13. Schematic representation of a polypeptide, a chain of many amino acids.

The long chain of amino acids then folds forming a three dimensional structure. The Figure 14 shows the schematic representation of a folded polypeptide. For simplicity, the folding shown in Figure 14 is absolutely random. In nature this is not really so, specific and ordered structures are observed in proteins.

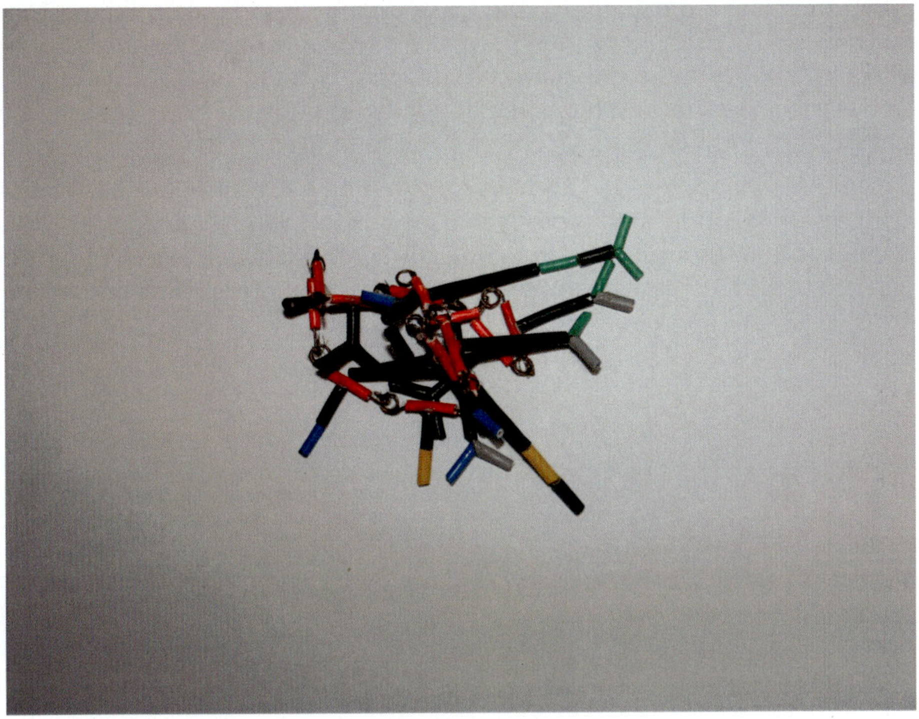

Legend of Figure 14. A folded polypeptide chain, which is a a folded chain of many amino acids.

In case of enzymatic proteins, the folded structure of proteins confers catalytic activity of enzymes that catalyse biochemical reactions. The sequence of amino acids in a protein, not just what amino acids are contained, determines the property of the enzyme. Sometimes two or more of such different proteins come together to form an enzymatic complex, which catalyses the biochemical reaction.

However, when we consume proteins for our nutrition, the sequence of the amino acids is inconsequential, as the digestive enzymes in our stomach catalyse cleavage of the chemical links between the amino acids and the free amino acids are absorbed in our intestine. Then new proteins of new sequences are formed in our system making use of these amino acids.

Annexure to Chapter 11

The N-end of an amino acid comprises an amino group that contains one nitrogen atom and two hydrogen atoms (HHN-), while the C-end comprises a carboxyl group that contains one carbon atom, two oxygen atoms and one hydrogen atom (-COOH)

During protein biosynthesis, the C - end (shown as hook in Figure 11) of the amino acid 1 reacts with the N - end (shown as ring in Figure 11) of the amino acid 2 to form the link between the amino acids 1 and 2. This link had been described by Hofmeister as the amide bond and had been described by Fischer as the peptide bond.

In 1901Fischer succeeded in chemically joining two molecules of the amino acid glycine by peptide bonds and created glycyl glycine (glycine – peptide bond – glycine). In 1907 he succeeded in chemically synthesizing an octodecapeptide, which can be termed as a small synthetic protein containing 18 amino acids.

12

BEADLE AND TATUM'S ONE GENE ONE ENZYME HYPOTHESIS

We have travelled for too long and through many countries, we might have lost track of some of the things that we have come across. Let us recapitulate. Archibald Edward Garrod had attributed the disease Alkaptonuria to a defective hereditary unit. Hereditary units had been given the name genes by Wilhelm Ludwig Johannsen. Garrod had guessed that the hereditary unit or the gene had the information about how to carry out a chemical reaction that converts one substance into another in a living object. Jöns Jacob Berzelius, who had discovered the role of catalysts in chemical reactions, had guessed that chemical reactions that go on in a living object, might also be mediated by catalysts. Wilhelm Kühne coined the term "enzymes" for the biological catalysts that function in living objects. Hence, a defective gene had resulted in a defective enzyme. This was the first ever inkling as to how a gene functions. James Batcheller Sumner had established that enzymes were protein in nature. Franz Hofmeister and Emil Fischer opined that proteins were linear arrays of many different amino acids linked by chemical bonds. So enzymes were very much real, just a chemical substance. The gene was still something hypothetical. Something very real was being synthesized under the guidance of something hypothetical. Nobody had any idea of the nature of the link between an enzyme and a gene.

In the fifth decade of the twentieth century, two persons were trying to understand the nature of this link. We have to travel to Stanford University in California, U.S.A. to peep into their activity. Edward Lawrie Tatum had teamed up with George Wells Beadle in the Department of Biology. They were aware that the defect in the hypothetical gene that results in Alkaptonuria in humans apparently occurred spontaneously in the first place, making the gene non-functional. Beadle and Tatum were wondering whether they could artificially create defects in a number of genes and trace the enzymes that might be affected. They had no clue as to what comprised a gene, but whatever it was, they would try to inactivate it by using radiation like X-ray and check if any enzyme has been inactivated. Obviously, they could not experiment with human beings. They needed a simpler organism that reproduced fast. They chose to work with *Neurospora*, the pink coloured mould that are seen to grow on stale bread. Indeed Beadle and Tatum discovered that each inactivated gene resulted in an inactive enzyme. Accordingly, Beadle and Tatum propounded their "one gene one enzyme" hypothesis in 1941, suggesting that a specific gene is responsible for the formation of a specific enzyme, which catalyses a specific biochemical reaction in a living organism.

George Wells Beadle, "Beets" to his close friends, was born in 1903 in Wahoo, Nebraska, U.S.A. He lost his mother when he was only four year old and was raised by his father and house keepers. His father, Chauncey Elmer Beadle was a farmer and grew potatoes, asparagus, strawberries, and kept bees, in his 40 acre farm. George was going to school, but would help his father in farming in his off time.

George Beadle

George's father wanted him to join him in farming whole time after finishing school. George's teacher in Wahoo High School, Bess MacDonald, however, recognized his talent and encouraged him to go to college. George Beadle joined the State University of Nebraska, College of Agriculture and obtained the degree of Bachelor of Science in 1926 and Master of Science in 1927. He then went to Cornell University and obtained the Ph.D. degree in 1931. He subsequently joined the Department of Biology in Stanford University, California, U.S.A. as a faculty.

Beadle's co-worker, Edward Lawrie Tatum was born in 1909 in Boulder, Colorado, U.S.A. His father was a learned person, Professor of Pharmacology. Edward did his Bachelor's degree in Chemistry in 1931, Master's degree in Microbiology in 1932 and Ph.D. degree in Biochemistry in 1934 from The University of Wisconsin. . He stayed on in Wisconsin as a postdoctoral fellow till 1937, in between working for a year in the University of Utrecht, Netherland.

In 1937 he saw a job advertisement, one George Beadle was looking for a Research Associate to help him in his pursuit in the new field of Biochemical Genetics.. He applied and got the offer. Meanwhile, Tatum had an offer from a butter company, which would have paid him more money. At that point, Arthur Tatum, Edward's father, made known to his son his apprehension that if he decided to work in an area which Arthur thought was nothing but a hybrid between Biochemistry and Genetics, "he would find himself an academic orphan, disowned by each of the disciplines of Biochemistry, Microbiology, and Genetics". In spite of all these, Tatum decided in favour of joining George Beadle as a Research Associate.

Tatum joined Beadle in the Department of Biology in Stanford University in 1937. Tatum's expertise in Microbiology and Biochemistry complemented Beadle's expertise in Genetics, resulting in a formidable combination.

Beadle and Tatum found that the mould *Neurospora*, could be grown in the laboratory in a medium containing the simple sugar glucose, common salt, a few other minerals and biotin, a B vitamin . Obviously, *Neurospora* could convert glucose into all the other essential constituents of its body, apparently through all kinds of biochemical reactions catalysed by specific enzymes.

Their logic was straight forward. Does not matter what a gene was, if it was damaged, the enzyme that particular gene was producing would be inactivated, hence the biochemical reaction that enzyme was catalysing, would cease to operate. There would be some chance that the affected biochemical reaction was producing in native *Neurospora* a compound that was an essential nutrient of *Neurospora*. So the irradiated *Neurospora* would fail to grow on the simple medium containing the sugar glucose, common salt, a few other minerals and biotin, but would resume growing if the essential nutrient that has ceased to be produced was also added as a supplement.

They knew that their research project was fraught with uncertainty, but nevertheless they were not the people to shy out. They decided that they would check at least 5,000 cultures that have been irradiated with X-ray before thinking of abandoning the idea. Fortune indeed favours the brave, they encountered a deficient *Neurospora,* that failed to grow on the minimal simple medium containing the sugar glucose, common salt, a few other minerals and biotin, in the 299th irradiated culture. The deficient *Neurospora* would revive and thrive if a new vitamin or an amino acid or some such compound was supplemented. Beadle and Tatum called these survivors deficient "mutants". One such mutant grew if the amino acid arginine was supplemented.

It is important to understand that if the radiation had directly damaged the enzyme, but not the gene, then its offspring would not require arginine, but would grow in the simple medium.. If one tears off the wing of a fly, its offspring would not be wingless, one has to knock off the gene(s) responsible for development of wings to get a wingless fly generation after generation. The mutant character of *Neurospora* persisted generation after generation, hence was hereditary because of the defect created in a gene.

Beadle and Tatum inferred that the biochemical reaction which led to the biosynthesis of the amino acid arginine, has been abolished in the arginine deficient mutants, because the gene of the specific enzyme that catalysed the specific biochemical reaction leading to the formation of arginine has been affected by the radiation of X-ray.

They subsequently isolated number of mutants deficient in other amino acids which could be revived by supplementing those particular amino acids.

The net finding then was that each different enzyme that catalysed the production of each different amino acid, had a separate gene. The novelty of the hypothesis, "One Gene One Enzyme", was that it would hold good irrespective of the fact that Beadle and Tatum did not know what exactly the gene consisted of. Though they had worked with only one organism, they very correctly extended their hypothesis to all organisms.

Before publishing their work in 1941, Beadle went to the California Institute of Technology to give a talk on their work. This is what Norman H. Horowitz, who later worked with Beadle as a post-doctoral fellow, had written about the talk. "Since the first Beadle-Tatum paper on *Neurospora* had yet to be published, no one in the audience had an inkling of what was to come.

The seminar was memorable. The talk lasted only half an hour, and when it was suddenly over, the room was silent. The silence was a form of tribute. The audience was thinking: Nobody with such a discovery could stop talking

---

Foot note: Annexure to Chapter 12 contains the biochemical pathway for the formation of arginine in Neurospora and identification of four different classes of arginine deficient mutants.

about it after just thirty minutes—there must be more. Superimposed on this thought was the realization that something historic had happened. Each one of us, I suspect, was mentally surveying, as best we could, the consequences of the revolution that had just taken place."

Horowitz subsequently wrote about his experience after he joined the laboratory of Beadle in Stanford University, "Every day brought unexpected new results, new mutants, new phenomena. It was a time when one went to work in the morning wondering what new excitement the day would bring. Beadle presided over this scientific paradise with the enthusiasm, intelligence, and good humour that characterized everything he did. He was a popular and much admired boss. He worked in the laboratory with everyone else. He especially enjoyed working with his hands, and he had plenty of opportunity to indulge himself in this regard. I came to work early one morning and found him painting one of the rooms. All this was in addition to his research and his teaching duties as a faculty of the Department of Biology. He always did more than anyone else".

In 1945 the same California Institute of Technology offered Beadle the Chairmanship of its Division of Biology and Beadle accepted it. In 1958 he shared the Nobel prize with Tatum. In 1961 Beadle was requested by the Board of Governors of the University of Chicago to join as the President of the University and he served there till his retirement in 1969.

In spite of his intense academic activities, George Beadle continued to work in his garden. The garden around his home in Stanford University produced enough corns and vegetables for two families. In the University of Chicago, he could be seen working in his garden early every morning. "Some were surprised to discover that this man was the president of the University, having thought him a hired gardener!" Beadle was an ardent skier and mountain climber, he had climbed California's Mount Whitney six times.

In 1981, Beadle developed Alzheimer disease, he died in 1989.

By 1941 Tatum and Beadle had completed their revolutionary work leading to the hypothesis, one gene one enzyme. The results of their work was published in the acclaimed journal, Proceedings of the National Academy of Sciences of U. S. A. Tatum was promoted to the position of Assistant Professor in Stanford University.

Edward Tatum

By 1945, however, academic politicking raised its head in the Biology Department in the wake of financial exigency. How can a chemist fit in a Biology Department? That was the question raised, as Tatum's Bachelor's degree was in Chemistry. In spite of all his academic success, Tatum was in distress. He remembered his father's foreboding that he may find himself an academic orphan.

In 1945, Tatum accepted an offer from Yale University to develop the Microbiology program there. That was the end of the glorious Beadle and Tatum era. Shortly afterwards, Beadle also left Stanford University and joined the California Institute of Technology at Pasadena. Interestingly, Tatum was invited back to Stanford in 1948, but in 1957 he left again to accept a Professorship in Rockefeller Institute in New York and stayed there until his death in 1975 from heart failure and emphysema because of cigarette smoking.

Annexure to Chapter 12

The biochemical pathway for the formation of arginine in *Neurospora* is the following:

Acetyl ornithine → Ornithine → Citrulline → Arginine

Beadle and Tatum argued that an arginine deficient mutant could arise as a result of inactivation of either of at least four different genes, (i) the gene for the enzyme that synthesizes acetyl ornithine, or (ii) the gene for the enzyme that catalyses the step, acetyl ornithine → ornithine, or (iii) the gene for the enzyme that catalyses the step, ornithine → citrulline, or (iv) the gene for the enzyme that catalyses the step, citrulline → arginine. Indeed they could classify the arginine deficient mutants into four classes.

Class 1. The mutants grew if either of acetyl ornithine, or ornithine, or citrulline or arginine was supplemented.

Suggestion: This class had a lesion in the gene for the enzyme that catalyses the synthesis of acetyl ornithine.

Class 2: The mutants grew if either of ornithine, or citrulline or arginine was supplemented, but did not grow if acetyl ornithine was supplemented.

Suggestion: This class had a lesion in the gene for the enzyme that catalyses the step, acetyl ornithine to ornithine.

Class 3: The mutants grew if either of citrulline or arginine was supplemented, but did not grow if acetyl ornithine or ornithine was supplemented.

Suggestion: This class had a lesion in the gene for the enzyme that catalyzes the step, ornithine to citrulline.

Class 4: The mutants grew if only arginine was supplemented, but did not grow if acetyl ornithine or ornithine or citrulline was supplemented.

Suggestion: This class had a lesion in the gene for the enzyme that catalyses the step, citrulline to arginine.

13

FREDERICK GRIFFITH'S OBSERVATION ON DIPLOCOCCUS PNEUMONIAE

We have been in Stanford University, U.S.A. busy admiring the path-breaking work of Beadle and Tatum being done with *Neurospora* mutants, but more than a decade earlier to the enunciation of the one gene one enzyme hypothesis, across the Atlantic Ocean, Microbiologist Frederick Griffith of a British government pathological laboratory in London was routinely looking under the microscope at the bacteria *Diplococcus pneumoniaee*, that caused pneumoniae. In 1928 he came across two distinct types, one having a smooth slimy coat forming large colonies that he named S form and the other having a rough coat with no slime around forming small colonies, that he named R form (Figure 15). The R form when injected into mice, did not do any harm, while the injected S form killed the mice.

Frederick Griffith

Frederick Griffith was born in the city of Hale in Cheshire, England in 1877. He graduated from Liverpool University in 1901 and then worked for the British government as Microbiologist in their pathological laboratories. It is in these laboratories that Griffith came across large number of *Diplococcus pneumoniae* strains collected from patients all over Britain. After the World War I, economy measures prevailed in Britain and the laboratories were starved of funds. In the language of an admirer, "Griffith and his colleague, William M. Scott could do more with a kerosene tin and a primus stove than most men could do with a palace".

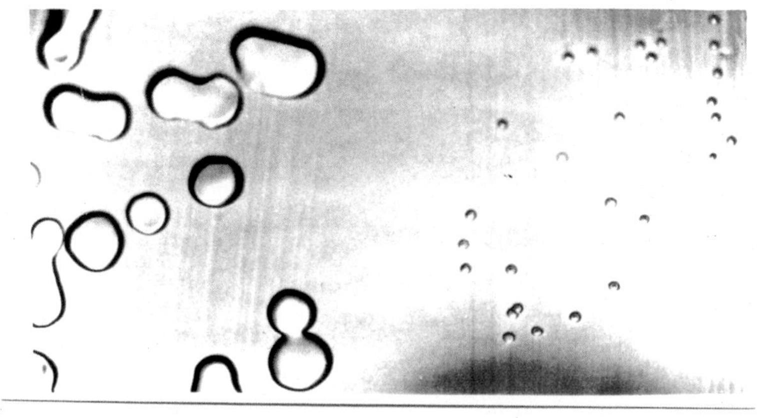

Legend of Figure 15. Photographs of colonies of R form (at the right) and S form (at the left) of *Diplococcus pneumoniae*. *[Reproduced from Avery, Macleod and McCarty, J Exp Med 1944; 79:17 by permission from The Rockefeller University Press.]*

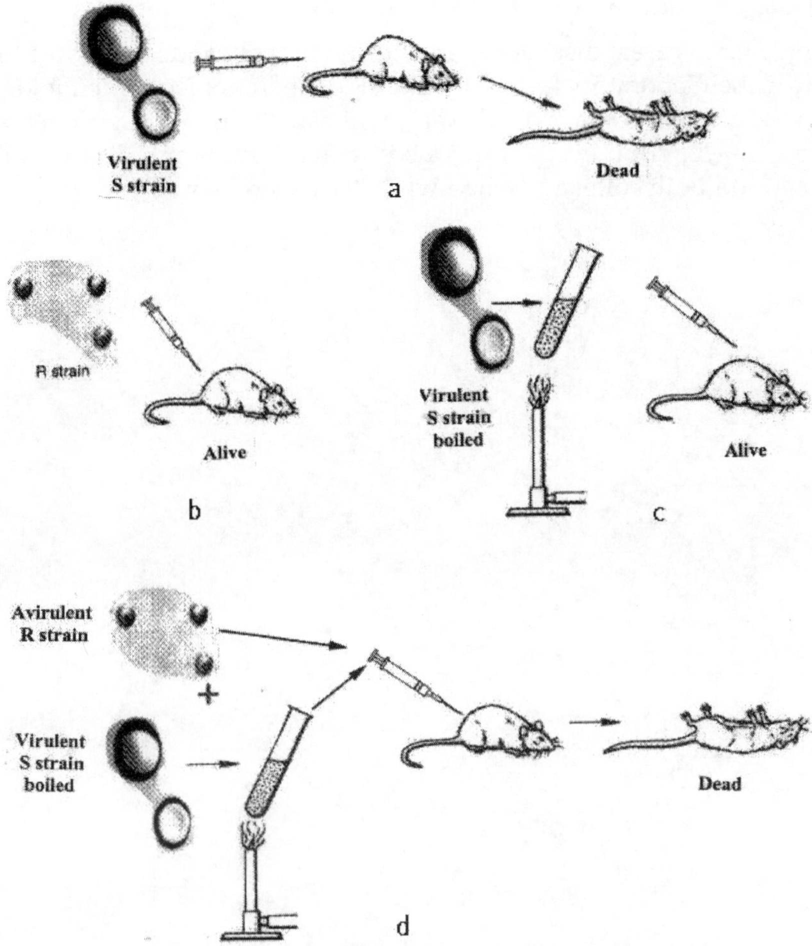

Legend of Figure 16. Effect of injection of *Diplococcus pneumoniae* into mouse. *(a)The mouse gets killed when the virulent S strain was injected, while (b) the mouse was unaffected on injection of the avirulent R strain. (c) Injection of the boiled S strain was harmless, but (d) injection of the boiled S strain mixed with the live R strain killed the mouse.*

[Adapted from Molecular Genetics by Stent and Calendar, 2nd edition, by permission,of Professor Gunther Stent]

The most Interesting observation of Griffith was that boiled and killed S form mixed with live R form, if injected into mice, killed them (Figure 16). The S form by itself, when killed by boiling and then injected into mice, failed to harm the mice. The blood of the mice killed by injecting the mixture of live R form and heat killed S form was full of the live S form. Griffith inferred that something from the dead S form has transformed the harmless R form into killer S form. The transforming principle of Griffith and gene of Johannsen and of Garrod and of Beadle and Tatum were possibly the same thing.

Even after his great discovery, Griffith and his colleague, Scott would work tirelessly in their laboratory to search patterns of pneumoniae epidemiology. This was in spite of the fact that World War II was raging and the German bombers were bombing London every day. One day in 1941 a bomb directly hit their laboratory and both colleagues died while they were working.

14

OSWALD AVERY'S DEMONSTRATION THAT THE TRANSFORMING PRINCIPLE OF GRIFFITH WAS NOTHING BUT DNA

Strangely enough, neither Griffith nor other workers of that time cared to purify or characterize the 'transforming principle'. Two decades later, across the Atlantic Ocean in the Rockefeller Institute Hospital in New York, Oswald Theodore Avery picked up the thread and purified the 'transforming principle'. Chemical tests revealed in 1944 that the 'transforming principle' was DNA. Since gene and the transforming principle were very likely the same thing, gene was nothing but DNA.

Oswald Theodore Avery was born in 1877 in Halifax, Nova Scotia, Canada. His father, Joseph Francis Avery was a pastor in the Baptist church. They moved to New York City when Avery was ten years old. Francis was invited to work as a pastor in the Baptist church in New York's lower east side. Avery was good at playing the cornet and would play it at the entrance of the church every Sunday afternoon to attract passers by to enter the church. He even got a scholarship to the National Conservatory of Music, but never used it. He chose to study Philosophy, English Literature, Political Economy, Public Speaking and History of Art in college and graduated in humanities in 1900, scoring 9 out of 10.

Oswald Theodore Avery

Avery then decided to change track completely and study medicine. He joined the College of Physicians and Surgeons of Columbia University in New York. He scored good grades there also. He completed study of medicine in 1904 and started general medical practice. Soon, however, he developed a feeling of helplessness, as he could not alleviate the sufferings of many of his patients. Microbial infections were difficult to treat those days and Avery decided to switch to research in the area of Medical Microbiology.

In 1907 he joined the Hoagland Laboratory in Brooklyn, New York as Assistant Director. His research work there on pathogenic bacteria attracted the attention of Dr. Rufus Cole, the Director of the Rockefeller Institute Hospital at Manhattan, New York, who invited Avery to join his Hospital as a Bacteriologist. In 1913, Avery joined the Rockefeller Institute Hospital and continued working with pathogenic bacteria including *Diplococcus pneumoniae*. He retired in 1943, when he was in his late sixties. He, however, continued working as an Emeritus Scientist in Rockefeller Institute Hospital and in 1944 he discovered that the transforming principle was DNA.

Avery repeated Griffith's experiment with the purified DNA. Indeed, the purified DNA from the S form of *Diplococcus pneumoniae,* when mixed with the harmless R form of *Diplococcus pneumoniae,* transformed some of the R form cells into the killer S form. The mixture of the harmless R form and purified DNA from the S form killed the mice when injected into the mice. As expected, the blood of the dead mice was again full of S form of the bacteria. Just the purified DNA from S form DNA did not kill the mice. Avery treated his 'transforming principle' with an enzyme that was known to degrade DNA. The transforming activity was lost. However, when he treated it with an enzyme that was known to degrade protein, the transforming activity was unaffected. It appeared that some toxic substance was produced based on information embedded in DNA that contributed to the lethality.

At that point of time the Australian scientist, Macfarlane Burnet happened to visit Avery's laboratory. It is said that Burnet was so excited that he immediately called his fiancée in Australia to tell her that Avery "has just made an extremely exciting discovery which, put rather crudely, is nothing less than the isolation of a pure gene in the form of DNA". One report in the renowned scientific journal *Nature* mentioned, "the genetic implications of this work are considerable". Geneticist Herman Muller wrote, "If this conclusion is accepted, their finding is revolutionary". Biochemist Howard Mueller wrote, "The importance of these observations can scarcely be overestimated". Geneticist Joshua Lederberg, who later own the Nobel prize, wrote in his diary: "Terrific and unlimited in its implications... I can see real case for excitement in this stuff". The X-ray crystallographer, William Astbury told a friend that the Avery group's finding was "one of the most remarkable discoveries of our time".

Avery, who was a bachelor, used to spend most of his time in the laboratory. He would rarely attend meetings or conferences. He would, however, not miss his yearly summer vacation, when he would enjoy sailing in Deer Island in the state of Maine in U.S.A.

Avery was awarded the Gold Medal of the New York Academy of Medicine and the Copley Medal by the Royal Society of London. He also received the Lasker award. As he was not awarded the Nobel prize, Nobel Laureate Arne Tiselius said that "Avery was the most conspicuous omission from the list of Nobel prize winners".

In 1948 he moved to Nashville and rented a house close to the house of his brother Roy. He would spend his time gardening and taking long walks. In the summer of 1954, when he was enjoying his favourite sailing trip in Deer Island, Avery felt acute pain in his abdomen. He was admitted to the hospital and had to undergo surgery, which revealed that he had fallen a victim of liver cancer. Avery died the next year at the age of 77.

15

FRIEDRICH MIESCHER HAD DISCOVERED DNA AS EARLY AS 1869

The work of Oswald Theodore Avery established that the 'transforming principle' was DNA, hence, the genetic material was nothing but DNA. DNA as a chemical substance was already known then, but its biological function was not known. In fact, DNA was purified from human pus cells in the University of Tubingen in Germany by Johannes Friedrich Miescher more than seven decades before Oswald Avery discovered that the transforming principle of Griffith or the genetic material was DNA. Miescher, however, had named the substance he had purified from pus cells as 'nuclein'.

Johannes Friedrich Miescher was born in 1844 in Basel, Switzerland in a family of medical doctors. Both his father and uncle were well known physicians and Professors of Anatomy and Physiology in the University of Basel. Friedrich was a shy but intelligent boy, who loved music. He wanted to be a priest, but his father did not approve of the idea. So he started studying medicine in Basel at the age of 17. In 1865 he was afflicted with typhoid fever. He recovered after long illness, but his hearing was impaired. He, nevertheless, graduated in medicine in 1867 when he was only 23.

Johannes Friedrich Miescher

Miescher was only 25 when he had purified 'nuclein'. He observed that 'nuclein' was not affected by enzymes that degraded proteins and fats. He further found that 'nuclein' contained the element phosphorus, but no sulphur. It was known then that all proteins contained sulphur, but no phosphorus. Subsequently, Miescher discovered that nuclein was present in sperms of quite a large number of animals. He wrote, "if … a specific substance is the cause of fertilization, one would without doubt have to think of the nuclein". He even speculated that nuclein might be involved in the "transmission of hereditary traits". It was subsequently revealed that Friedrich Miescher's 'nuclein' and DNA were the same thing.

Friedrich Miescher devoted his life to medical research, as medical practice would be difficult with his impaired hearing. He became the Professor of Physiology at the University of Basel in 1872, a position previously held by his father and then his uncle. As a Professor he had to teach, but he was not a good teacher, may be because he was quite shy. He, however, continued to work very long hours in the laboratory. He established Switzerland's first Physiological Institute in 1985. He died of tuberculosis in 1895 at the age of 51. A laboratory of the Max Planck Society in Tübingen and a Research Institute in Basel have been named after him.

16

WHAT IS THE CHEMICAL NATURE OF DNA?

Now that DNA has occupied the centre stage, let us look at the chemical nature of DNA. DNA is the abbreviation of deoxyribonucleic acid. DNA is a long linear molecule comprising a chain of alternating phosphate (P) and deoxyribose sugar (dR). To the deoxyribose units are hooked the purine bases [adenine (A) or guanine (G)] or the pyrimidine bases [cytosine (C) or thymine (T)]. The size of G and A is about the same, while the size of T and C is about the same. The size of G and A is about double the size of T and C.

We can compare the DNA chain with a snake. No, please don't panic or shake with fear. The other day our son discovered a snake in his ground floor apartment. He informed the Wild Life Protection Society of India. The person from the Society looked at the snake and immediately pronounced that the snake was not poisonous. He picked up the snake with his bare hand and put it in his bag.

The snake that I am talking about is also not poisonous, though like other snakes it has a head, a tail, a belly and a back. The unique thing about this snake is that it has alternate black and white bands on its body. The black band is analogous to the phosphate (P) moiety of DNA, while the white band of the snake is analogous to the deoxyribose sugar (dR) moiety. Next to the white band there is a black band, then again a white band, then again a black band, then again a white band and so on. Finally the tail of the snake is there which symbolises the 3' end of DNA.

The snake is lying on its belly. The loads marked A, G, C, and T on the back of the snake on its white bands represent the purine bases [adenine (A) or guanine (G)] and pyrimidine bases [cytosine (C) or thymine (T)]. hooked on to the deoxyribose units of DNA. The left end of the small segment of DNA shown in Figure 17 is the 5' end and the right end is the 3' end. The head of the snake represents the 5' end of DNA and the tail of the snake represents the 3' end.

The height of the load (G) and of the load (A) on the back of the snake on its white bands is about the same, while the height of load (T) and of the load (C) is about the same. The height of load (G) and of the load (A) is about double the height of the load (T) and of the load (C). Figure 17 shows a very small representative segment of DNA. The sequence A,C,G,T shown in Figure 17 is arbitrary. In nature, the sequence is random.

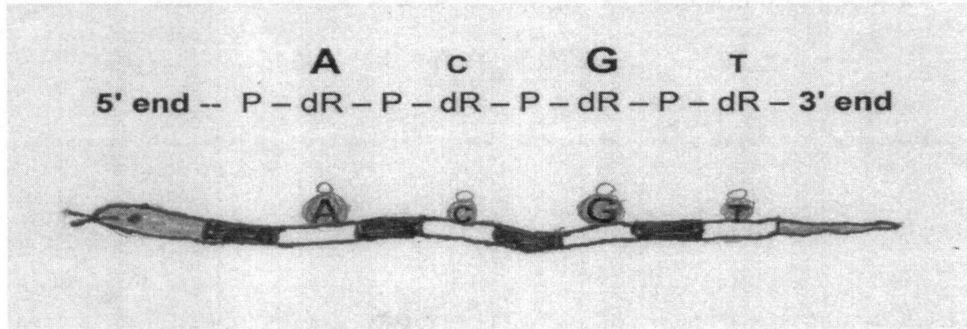

Legend of Figure 17. Analogy of a segment of DNA (top element) with a snake (bottom element)..

Foot note: The Annexure of Chapter 16 contains the chemical representation of a short fragment of the linear DNA molecule and the structures of the purines adenine (A) and guanine (G) and the pyrimidines cytosine (C) and thymine (T).

Annexure of Chapter 16

Following Figure 18 is a schematic chemical representation of a short fragment of the linear DNA molecule and the structures of the purines adenine (A) and guanine (G) and the pyrimidines cytosine (C) and thymine (T).

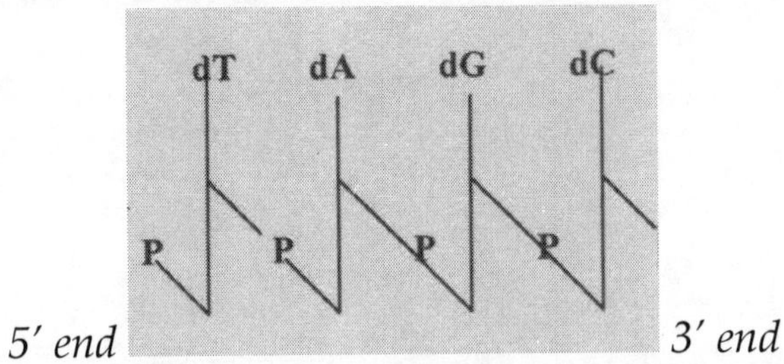

Legend of Figure 18. Schematic chemical representation of a short fragment of a DNA chain. *The vertical lines represent the deoxyribose moieties. Ps represent the phosphate groups that link the deoxyribose moieties. The fifth carbon atom of one deoxyribose is linked by the phosphate group to the third carbon atom of the deoxyribose immediately left of it. The phosphate (P) at the left end is the 5' end of the DNA fragment, while the hydroxyl (not shown) at the right end is the 3' end. The purines and the pyrimidines are linked to the first carbon atoms of all the deoxyribose moieties. dT represents the pyrimidine thymine and dC represents the other pyrimidine cytosine, while dA represents the purine adenine and dG represents the other purine guanine of DNA.*

Purines comprise two cyclic groups fused together, while pyrimidines comprise a single cyclic group. The purines of DNA are adenine (A) and guanine (G), while the pyrimidines are thymine (T) and cytosine (C). The size of G and A is about the same, while the size of T and A is about the same. The size of G and A is about double the size of T and C. The chemical structures of adenine (A), guanine (G), thymine (T) and cytosine (C) are shown in Figure 19 below just to give a comparative idea of their sizes.

Legend of Figure 19. Chemical structures shown just to give a comparative idea of the sizes of the purines and pyrimidines that are constituents of DNA. *From left to right: the purines, adenine (A) and guanine (G); then the pyrimidines, thymine (T) and cytosine (C).*

17

ERWIN CHARGAFF DETERMINED THE AMOUNTS OF DIFFERENT BASES OF DNA

Oswald Theodore Avery did not get the Nobel prize. Some people believe that this was because many of the stalwarts in the field still believed that proteins comprised the genetic material, as DNA with just four bases [the two purine bases adenine (A) and guanine (G) and the two pyrimidine bases thymine (T) and cytosine (C)] was unlikely to provide the variety needed. Besides, there was the feeling that the proportion of bases in DNA was the same irrespective of the organism from which the DNA was isolated. If that was so, then there would possibly be no difference between DNA isolated from let us say man and DNA isolated from cow or DNA isolated from the mold *Neurospora* or the DNA isolated from the bacteria *Diplococcus pneumoniae*. Then how can DNA be the genetic material?

We had to wait for about a decade after Avery's discovery, when we came across a person named Erwin Chargaff, who was then an Assistant Professor in the Department of Biochemistry of Columbia University in New York city, who was trying to look at DNA more closely.

Erwin Chargaff was born in 1905 in a wealthy family in the town of Czernowitz, capital of Bukovina, the easternmost province of the then Austro-Hungarian empire. His father was Herman Chargaff, a well educated German speaking Austrian Jew. The Chargaff family were enjoying vacation in a Baltic sea resort in 1914, when the World War I broke out. As the news of imminent occupation of Czernowitz by the Russian army reached them, Herman Chargaff abandoned the idea of returning to Czernowitz and decided instead to go to Vienna, leaving behind all his properties in Czenowitz.

The Chargaffs had a small library at home and that was the catalyst for the child Erwin to develop his fascination for literature. Erwin went to one of the best schools in Vienna and studied Greek and Latin there. He excelled in both the languages. He immensely enjoyed reading Western classical literature. His mother Rosa, who was also well educated, presented Erwin the complete works of Goethe, when Erwin was only 15 years old. He wrote in his diary, "My real reading began, I believe, in 1920, when my mother gave me as a present Goethe's complete works in the sixteen volumes of the beautiful Insel edition. They still stand much read on my shelves".

Unlike other children of his age, Erwin's evenings in Vienna were spent with his friend Albert Fuchs, either discussing European classical literature and poetry, or listening to compositions of Mozart and other European composures. His most admired writer was Karl Kraus and he would not miss recitals by Karl Kraus. Erwin himself was destined to be a writer.

Erwin Chargaff

The devastating inflation of 1923, however, was a big blow to his father and Erwin was forced to choose a career of a Chemist that would ensure him employment in the Chemical industry. He studied Chemistry for five years in the University of Vienna and obtained his Ph.D. degree in 1928. By then unemployment was high in Vienna, very few jobs but too many graduates. There were riots in the streets of Vienna in 1927.

Erwin Chargaff decided to try his luck in U.S.A. He applied for the Milton Campbell Research Fellowship in Organic Chemistry of the Yale University. His good Ph.D. work was appreciated and he was offered the fellowship. He did post-doctoral work for two years with Professor R. Anderson, an editor of the prestigious Journal of Biological Chemistry. Anderson was impressed with Chargaff's work and wanted him to stay on U.S.A. Chargaff, however, was keen to return to Europe. As he had mentioned in his autobiography, "the endless commotion of New York city and Yale's caste provincialism" were disgusting.

He found a job in the Public Health Department in the University of Berlin in 1930. For two and a half years he enjoyed "the best years of his life" as he immersed himself in the lively cultural life of Berlin. In the University he impressed everybody, not only with his skill in Analytical Chemistry, but also with his proficiency in Central European literature. In 1933, however, the scene changed completely, as Adolf Hitler became the Chancellor of Germany. Chargaff, being a Jew, realized that he had no place in Germany.

He moved to Paris on invitation from Albert Calmette, Director of Institut Pasteur. Chargaff, however, did not enjoy being in Paris as he found it "decadent". He tried, but failed to get a job in U.K. Finally, he decided to go to U.S.A. again. He found a research position in the Department of Medicine in the College of Physicians and Surgeons in Columbia University in New York. The scientists of U.S.A. at that point of time, were welcoming with open arms, all eminent refugee scientists from Europe.

In 1938, Hans Clarke invited Chargaff to join his Department of Biochemistry in Columbia University as an Assistant Professor, as he recognized Chargaff's past work.

While in Columbia University, Chargaff developed a method of accurately determining the amounts of the different purine and the pyrimidine bases (A, G, T, C) of DNA and in 1950 discovered that the proportion of a particular base relative to all the other bases was quite different in DNA from man, cow, yeast and avian tuberculosis bacilli. This was a revelation and implied that DNA of each species was quite unique, very much in accordance with the view that DNA was the genetic material.

Chargaff made another striking discovery. He found that the proportion of adenine (A) was about the same as that of thymine (T) and the proportion of guanine (G) was about the same as that of cytosine (C) in all the four species studied by him (man, cow, yeast and avian tuberculosis bacilli). The implication of the second discovery was not obvious to Chargaff or other scientists of the day. Chargaff wrote in his diary, "When I began to realize how unique were the regularities we had discovered, I tried, of course, to understand what it all meant, but did not get very far. I attempted to build molecular models of the nucleotides… I ran out of atoms and even more my patience…. thus I missed the opportunity of being enshrined in the various halls of fame of the science museums".

Around that time Hitler took control of Austria. Chargaff tried his best to bring his mother to U.S.A., as his father had already died. He was unsuccessful and he never again heard from his mother. This was a matter of utter dismay to him for the rest of his life.

Chargaff remained in Columbia till his retirement in 1974 as full Professor and Head of the Department of Biochemistry. He was then made an Emeritus Professor and he continued his research activity till 1992. Though Chargaff's work laid the foundation of the new branch of science, Molecular Biology, he would prefer to call himself a Cell Chemist.

Chargaff became an American citizen in 1940, but he never forgot his mother tongue and was a prolific writer and a frequent radio and TV interviewee in German. He was a fellow of the National Academy of Sciences of U.S.A.. He won the Pasteur medal, the Charles Leopold Mayer prize and the Gregor Mendel medal. Till his death at the age of 96 in 2002, Chargaff enjoyed reading European classical literature in his apartment's library.

18

IN A BID TO UNDERSTAND THE THREE DIMENSIONAL STRUCTURE OF DNA, ROSALIND FRANKLIN WAS SUCCESSFUL IN DETERMINING ITS X-RAY DIFFRACTION PATTERN

While we were in New York, watching Erwin Chargaff trying to understand DNA by determining the proportion of the purine and pyrimidine bases adenine, guanine, thymine and cytosine in it, across the Atlantic Ocean in the city of London, a young lady named Rosalind Franklin was contemplating to make use of an altogether different approach to understand DNA.

Rosalind Elsie Franklin was born in 1920 in a prominent British Jewish family. Both her parents were well known for their charities and various community services. Rosalind went to St. Paul's School for Girls, "which emphasized preparing its graduates for careers, not just for marriage". In school she was good in science, mathematics and languages like French, Italian and German. She passed her matriculation in 1938 with six distinctions, winning a scholarship for three years for studying in the University. She donated her scholarship to a refugee student and joined Newnham College at Cambridge University. Rosalind obtained her Bachelor's degree majoring in Physical Chemistry in 1941. She then joined the Ph.D. programme in Cambridge, published five papers and obtained her Ph.D. in 1945.

Rosalind Elsie Franklin

During the war years, Rosalind volunteered to serve as an Air Raid Warden and patrolled around her designated area during air raids, looking after the needs and welfare of the people at great personal risk.

Rosalind Franklin had joined the British Medical Research Council's Biophysics Unit in King's College, London as a Research Associate in January 1951. She had since 1947 worked with Jacques Mering at the Laboratoire Central des Services Chimiques de l'État in Paris, where she had learnt the practical aspects of applying X-ray diffraction in elucidation of structures of amorphous and crystalline substances.

X-rays get diffracted or scattered when these fall on substances. The pattern of diffraction of X-rays reveals the arrangement of atoms in the substances. Because of her past experience, John Randall, the Director of the Biophysics unit, asked Rosalind Franklin to work on DNA fibers independently, even if her position was of a Research Associate. Randall felt that elucidation of the three dimensional structure of DNA would reveal how it functions. He also asked her to supervise the work of Ph.D. student Raymond Gosling, who was earlier working with Maurice Wilkins, another scientist of the Unit. This had upset Wilkins very much, as he had been working in the Unit on X-ray diffraction since 1950 and had expected Franklin to work under him. As a result, the relationship between Franklin and Wilkins had not been cordial right from the beginning.

In November 1951, Franklin presented the data she had generated with Gosling in a lecture that she delivered in King's College, London. She presented suggestive evidence that DNA was likely to have a closely packed helical structure containing two coaxial opposite chains, having the purines and pyrimidines inside and the phosphate-ribose chains outside. The double helix was thus first conceived by Rosalind Franklin. Figure 20 shows a short fragment of DNA as suggested by Franklin, except that for ease of drawing, the helix has been replaced by a straight duplex.. The corresponding snake analogy, as described in Chapter 16, can also be seen in Figure 20.

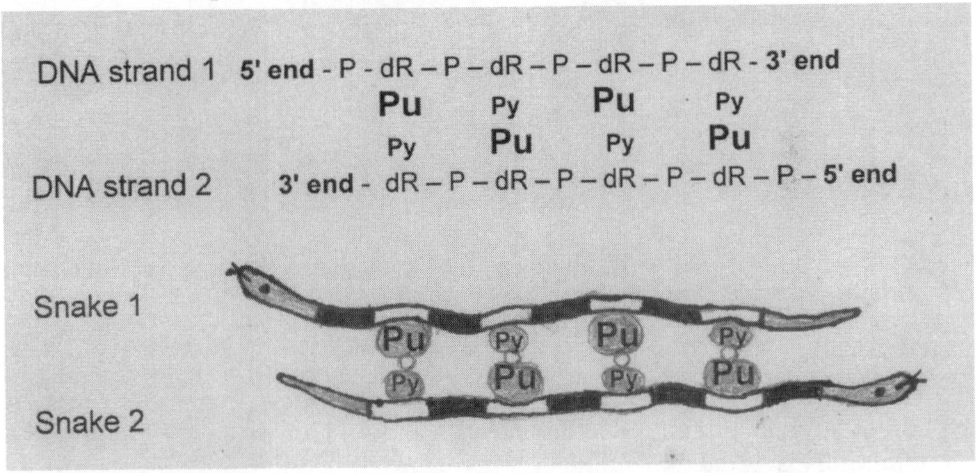

Legend of Figure 20: A short segment of duplex DNA as suggested by Rosalind Franklin. *Orientation of the two strands of DNA is opposite, the 5' end of DNA strand 1 and the 3′ end of DNA strand 2 are in one side, while the 3' end of strand 1 and the 5' end of strand 2 are in the other side. The purines (Pu) and pyrimidines (Py) are inside, while the deoxyribose and phosphates are outside. The purines of one strand are across the pyrimidines of the other strand and vice versa. This arrangement allows the duplex to have an uniform diameter, as the size of a purine is about double the size of a pyrimidine . Franklin's X-ray analysis could not distinguish between the two purines, adenine and guanine or between the two pyrimidines, thymine and cytosine.*

The snake analogy. *The head of the snake 1 and the tail of snake 2 are in one side, while the tail of snake 1 and the head of snake 2 are in the other side. The loads comprising the purines and pyrimidines are on the backs of the snakes. The load of a pyrimidine of one snake is across the load of a purine of the other snake and vice versa. The two snakes are in touch with each other through their loads on their backs, not through their bellies.*

Franklin had subsequently obtained her decisive and famous Photo 51 (Figure 21) which confirmed her earlier suggestion that DNA had a closely packed helical structure containing two coaxial opposite chains, having the purines and pyrimidines inside and the phosphate-ribose chains outside.

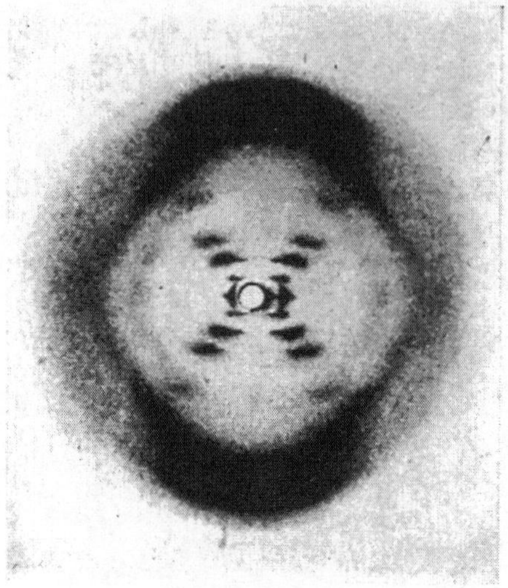

Legend of Figure 21 Photo 51, photograph of X ray diffraction pattern of DNA, form B. *[Reproduced from Franklin and Gosling, Nature 1953, 171: 740 with permission from Nature]*

Franklin was never in a haste to publish or build models without irrefutable experimental data, but she was the one who should have been given the credit for obtaining experimental evidence suggesting the double helical structure of DNA.

Science was everything to Rosalind Franklin. She had written, "Science and everyday life cannot and should not be separated". During academic discussions, Rosalind was always concise and direct and she would look straight at the discussant's eye. On the other hand, she would be very impatient with her student. Because of her conflict with Maurice Wilkins, Rosalind Franklin decided to leave King's College, and she moved to Birkbeck College of the University of London around the middle of March,1953. She was allowed to leave on condition that she would not continue working on DNA.

In Birkbeck, Franklin revealed the structure of a whole virus, the Tobacco Mosaic Virus, which had RNA as its genetic material. Subsequently she extended her work to Cucumber Virus 4, Turnip Yellow Mosaic Virus and other RNA containing plant viruses that affected potato, tomato and pea. In 1957, Franklin initiated work with Polio Virus, but could not complete because of her ill health.

Foot note: Detailed analysis of photo 51 can be found in Annexure of Chapter 18.

Around the middle of 1956, Franklin felt that her tummy had bulged. As she consulted her physician, she asked whether Franklin was pregnant. She answered, "I wish I were". Surgery was subsequently conducted, which revealed malignant tumors in her ovary. The tumors were removed and Franklin started recovering. She went back to the laboratory and resumed guiding the members of her group. They published seven papers in 1956 and six in 1957. Around the end of 1957, however, Franklin again fell ill, the cancer had recurred. As she felt better after treatment, she resumed working in January, 1958. She fell ill again by the end of March, and died on April 16, 1958 at the age of 37.

Her associates, Aaron Klug and John Finch published the Polio Virus structure the following year, dedicating the paper to her memory. Klug was awarded the 1982 Nobel Prize for this work. John Desmond Bernal said around the time of her death, "As a scientist Miss Franklin was distinguished by extreme clarity and perfection in everything she undertook." Jacques Mering wept when he visited her during her last days and mentioned that he was captivated by her intelligence and beauty.

Rosalind Franklin was a spinster. Evi Ellis, the war time child refugee from Europe, who had shared Rosalind's bedroom, tried match making, but did not succeed.

The July 21, 2020 issue of the reputed scientific journal Nature published an editorial article on the birth centenary of Franklin entitled, "Rosalind Franklin was so much more than the wronged heroine of DNA". The concluding line of the article said, "Franklin is mostly remembered for not receiving full credit for her contributions to the discovery of DNA's structure. That part of Franklin's life story must never be forgotten, but she was so much more than the 'wronged heroine', and it's time to recognize her for the full breadth and depth of her research career"

Annexure of Chapter 18

Following are detailed analysis of photo 51 by Rosalind Franklin.

1. The X-like pattern formed by the layer lines immediately confirmed a helical structure of DNA.
2. The two 3.4 Å reflections on the meridian corresponded to the inter-nucleotide distance along the fibre axis.
3. The intensity of the reflection at the meridian was, however, too strong for a single helix, so a double helix was obvious.
4. The 34 Å distance between the layer lines suggested that the nucleotides repeated ten times before the configuration repeated. Thus "there were exactly 10 residues per turn of the helix".
5. The diameter of the DNA double helix was calculated from the angle formed by the layer lines to be 20 Å.

6. "If ten phosphorus atoms lie on one turn of a helix of radius 10 Å, then the distance between neighbouring phosphorus atoms in a molecule is 7.1 Å. This corresponds to the P…P distance in a fully extended molecule, and therefore provides indication that the phosphates lie on the outside of the structural unit. The sugar and base groups must accordingly be turned inwards towards the helical axis"
7. "The relative distances of the repetitive elements in the DNA molecule, and the dimensions of what is called the monoclinic unit cell – which indicated that the molecule was in two matching parts, running in opposite directions".
8. Rosalind Franklin also realized "that the way the component nucleotides or bases on each strand were connected meant that the two strands were complementary, enabling the molecule to replicate". Franklin also foresaw that "an infinite variety of nucleotide sequences would be possible to explain the biological specificity of DNA", which clearly showed "that she had glimpsed the most decisive secret of DNA: the sequence of bases contained the genetic code".

19

JAMES WATSON AND FRANCIS CRICK CONSTRUCTED A MOLECULAR MODEL OF THE DOUBLE HELICAL DNA

From Birkbeck College of the University of London, we drove to Cambridge University. It was February 28, 1953, a Saturday. By the time we reached Cambridge University, it was lunch time . We went to the pub The Eagle for some food and drink. The pub was crowded. We heard a person saying something loudly. The person declared that they have "found the secret of life", the structure of DNA. The crowd was electrified. The person said, he and James Watson were successful in constructing a physical scale model of DNA revealing its double helical nature. We found out the name of the person was Francis Crick.

Francis Harry Compton Crick was born in 1916 in a small village near the English town of Northampton. He used to go to the village church with his parents, but at the age of 12 he started asking for scientific answers instead of accepting blind religious belief. Francis obtained the Bachelor's degree in Physics from University College, London, but his Ph.D. work was disrupted when a bomb fell through the roof of his laboratory during World war II. He joined the Cavendish Laboratory in 1949 to complete his Ph.D.

Francis Harry Compton Crick

Francis Crick's co-worker, James Dewey Watson was born in Chicago in 1928. He was a bright and inquisitive child, who would always ask, why. At the age of 15, he won a fellowship of "the gifted youngster program" and enrolled in the University of Chicago. Watson graduated from the University of Chicago in 1947. He received a fellowship from Indiana University for doing Ph.D. and obtained his Ph.D. degree there in 1950. He was only 22 years old then. Watson moved to the Cavendish Laboratory in Cambridge, England in October 1951, There he met Francis Crick and worked together for solving the structure of DNA. Crick was 12 years older and was still a Ph.D. student, while Watson already had a Ph.D., but the two "hit it off immediately".

James Dewey Watson

Watson and Crick were keen to solve the molecular structure of DNA, but they did not plan to carry out any experiment. Their approach was to construct a physical scale model based on intelligent analysis of results of other people's experiments. Watson had attended the November 1951 lecture of Rosalind Franklin in King's College. Watson did not follow all the details of the X-ray crystallographic evidence that Franklin had presented, but he understood that DNA was very likely a helical duplex. In early 1952, Watson and Crick constructed a physical model of double stranded DNA that had the deoxyribose – phosphate chain inside and the purine and pyrimidine bases outside. Their logic was that the deoxyribose – phosphate chain was just an inert scaffold, while the actual reactive elements were the purines and pyrimidines, so these should be outside. Besides, they imagined that the orientation of the two strands of DNA is similar, but not opposite. A representation of this first model of Watson and Crick, that was conceived in 1952, can be seen in Figure 22., except that for ease of drawing, the helix has been replaced by a straight duplex. Figure 22 also includes the image of the snake analogy described in chapter 16, according to the proposition made by Watson and Crick in 1952.

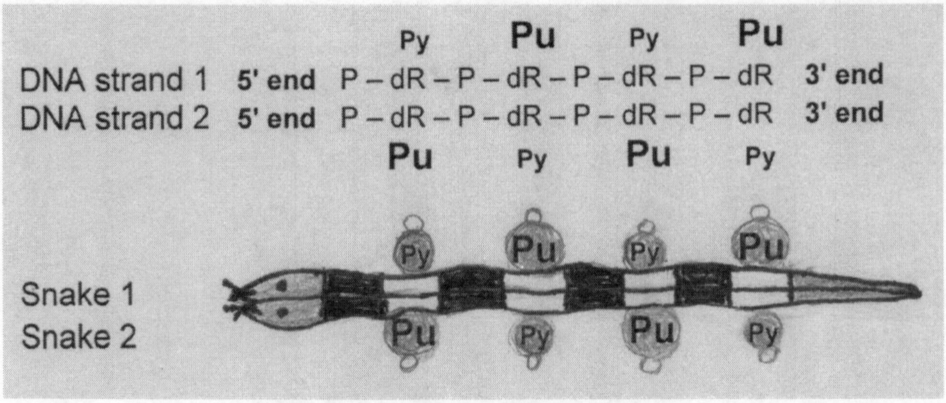

Legend of Figure 22.: *A short segment of duplex DNA as suggested by Watson and Crick in 1952. Orientation of the two strands of DNA is similar, the 5' end of DNA strand 1 and the 5' end of DNA strand 2 are in one side, while the 3' end of strand 1 and the 3' end of strand 2 are in the other side. The purines (Pu) and pyrimidines (Py) are outside, while the deoxyribose and phosphates are inside.* **The snake analogy.** *The head of one snake is in touch with the head of the other snake. The purine and pyrimidine loads are on the backs of the snakes. The two snakes embrace each other through their bellies.*

Meanwhile, the American chemist Linus Pauling, who discovered helical structure of fibrous proteins, had joined the race for finding the structure of DNA. Alarmed, Watson went to King's College on January 30, 1953 pleading for collaboration, so that they could solve the DNA structure together before Pauling could do it. But Watson, as young and brash he was, picked up an argument with Rosalind Franklin. Franklin said she was not in favour of building models without irrefutable experimental data. In reply Watson said Franklin did not know how to interpret her own data. Franklin was furious and there was a commotion. Maurice Wilkins intervened and took Watson to his office. There he showed Franklin's photo 51 to Watson, without caring for the simple ethics of seeking permission from Franklin. As Watson recalled later, "that when he saw the photo – which was far clearer than any other he had seen – 'my mouth fell open and my pulse began to race'."

Watson went back to Cambridge and described to Crick what he saw in photo 51. Max Perutz, who was the Ph.D. thesis adviser of Crick, had a copy of the report that was submitted by the Biophysics unit of Kings College to the review committee of the British Medical Research Council. The British Medical Research Council had been financially supporting the research work being done in the Biophysics unit of Kings College and the Council had sent a Review committee in December 1952 to to the Biophysics unit to assess their research work. Among other things, the report also contained all the detailed crystallographic calculations done by Rosalind Franklin to arrive at her conclusion that DNA had a double helical structure. Around the middle of February 1953, Perutz passed on that report to Crick. The photo 51 and Franklin's crystallographic calculations conveyed clearly to Watson and Crick that DNA had two strands. Also the deoxyribose – phosphate chain was outside and the purine and the pyrimidine bases were inside, not *vice versa,* in contrast to what they had conceived earlier. These also conveyed to them that the two strands of DNA ran in opposite directions, forming a closely packed helical structure of uniform diameter.

Franklin's photo 51 and crystallographic calculations did not, however, tell anything about what held the two DNA chains together. Watson and Crick appreciated that the bonds that held the two chains together should be weak bonds which could be easily broken and easily remade for reasons explained later in this chapter. They suggested hydrogen bonds would be appropriate for this purpose, which may be imagined to be similar to *'Velcro'*.

Franklin's photo 51 and crystallographic calculations did not also tell anything about which pyrimidine base lay across which purine base. Watson and Crick took the cue from Erwin Chargaff's finding on the proportion of bases in DNA, irrespective of its source (Chapter 17). They argued, if the proportion of adenine and thymine is the same in a duplex linear molecule, the simplest configuration would be that wherever there is an adenine residue in one linear polymer strand, there is a thymine residue right across it in the other polymer strand and *vice versa.* . Similarly, if the proportion of guanine and cytosine is the same in a duplex linear molecule, the simplest configuration would be that wherever there is a guanine residue in one linear polymer strand, there is a cytosine residue right across it in the other polymer strand and *vice versa.* Incorporating all these facts and ideas, Watson and Crick physically constructed a scale model of the DNA double helix. The following figure is a reproduction of the photograph showing Watson and Crick appreciating their model.

Legend of Figure 23: James D Watson and Francis HC Crick looking at their physical scale model of DNA. *[Reproduced from the Double Helix, Atheneum, NY, p 215 (Copyright 1968 by JD Watson. Photograph by AC Barrington Brown) by permission of Professor JD Watson.]*

A representation of the second model of Watson and Crick, that was conceived in 1953, can be seen in Figure 24, except that for ease of drawing, the helix has been replaced by a straight duplex. Figure 24 also includes the image of the snake analogy described in chapter 16 according to the proposition made by Watson and Crick in 1953.

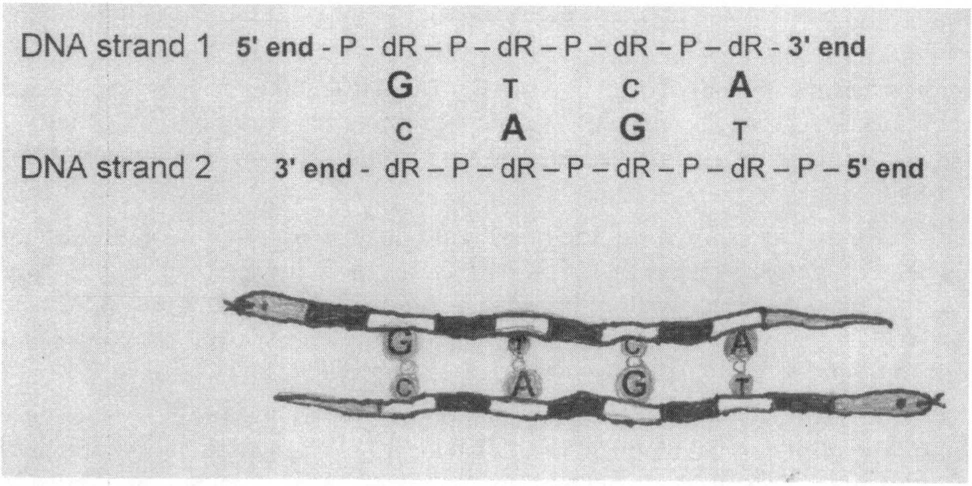

Legend of Figure 24. A short segment of duplex DNA as suggested by Watson and Crick in 1953. *Orientation of the two strands of DNA is opposite, the 5' end of DNA strand 1 and the 3' end of DNA strand 2 are in one side, while the 3' end of strand 1 and the 5' end of strand 2 are in the other side. The purines (A and G) and pyrimidines (T and C) are inside, while the deoxyribose and phosphates are outside. G in one DNA strand is across C in the other DNA strand, while A in one strand is across T in the other strand. This arrangement allows the duplex DNA to have an uniform diameter, in spite of the fact that the sizes of A and G are about double the sizes of T and C .*

The snake analogy. *The head of the snake at the top (snake 1) and the tail of the snake at the bottom (snake 2) are in one side, while the tail of the snake at the top (snake 1) and the head of the snake at the bottom (snake 2) are in the other side. The loads comprising the purines (A and G) and pyrimidines (T and C) are on the backs of the snakes. The load G of one snake is across the load C of the other snake and the load A of one snake is across the load T of the other snake.*

On February 28, 1953, the day Crick declared in the Cambridge pub that they have "found the secret of life", the structure of DNA, Crick told his wife after returning home in the evening, "we seemed to have made a big discovery." As he recalled, "years later she told me that she hadn't believed a word of it." As he recounted her words, "You were always coming home and saying things like that, so naturally I thought nothing of it ". On March 19, 1953, Crick wrote a seven page long letter to his son, who was in a boarding school, explaining their discovery. The letter began, *"My Dear Michael, Jim Watson and I have probably made a most important discovery..."*. That letter fetched U.S.$ 6,059,750, when it was put up for auction after Crick's death.

On March 8, 1953 Watson received a letter from Wilkins stating gleefully that, "Franklin was finally leaving and they could put all hands to the pump". Watson and Crick never revealed to anybody that they had obtained crucial clues

Foot note: Some details about the model can be seen in the Annexure of chapter 19

for building their model from Franklin's photo 51 and her crystallographic calculations. There was no clear acknowledgement to Franklin in their article in the scientific journal Nature of April 25, 1953 either, in which they published their model. There was only a clever and evading footnote which said, "having been stimulated by a general knowledge of Franklin and Wilkins' unpublished contribution".

In spite of all these, Rosalind Franklin did not bear any feeling of resentment or ill will against Watson or Crick. The archives of the Cold Spring Harbor Laboratory has a letter written by Rosalind Franklin informing James Watson about her later experiments with Tobacco Mosaic Virus which began with "Dear Jim" and ended with "Best Wishes, Yours, Rosalind".

We now look at some interesting facts about DNA. DNA molecules are very long. The total genetic material of the bacteria *Escherichia coli*, which inhabits our gut, is a single molecule of DNA about 4.6 million base pairs long. Figure 25 shows the photograph taken through an electron microscope of DNA released on lysis of an *Escherichia coli* cell.

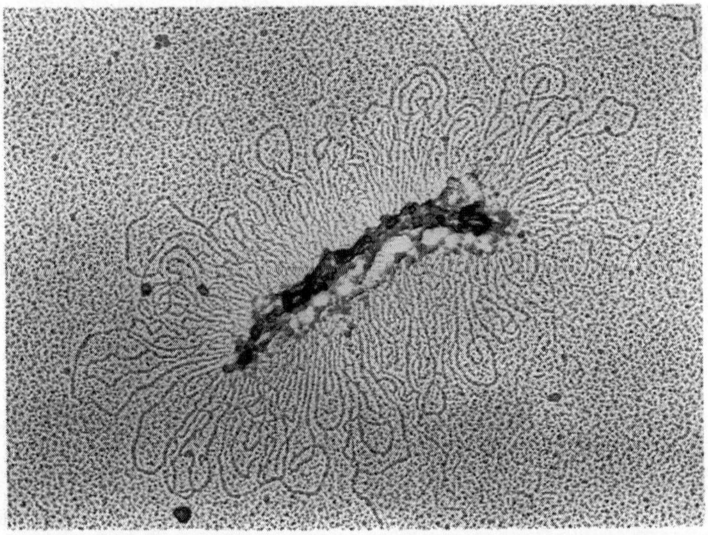

Legend of Figure 25. A photograph through electron microscope of a cell of the human gut bacteria Escherichia coli which has been lysed by using detergents. *The DNA, which is a single molecule, can be seen strewn all around the remnants of the lysed cell, that is in the centre. The DNA was stained with uranyl acetate. The photograph was taken in the laboratory of Prof. H. K. Das by Dr. Nirupam Roy Choudhury.*

A human cell contains 23 molecules of DNA. The longest one is about 249 million base pairs long, while the shortest one is about 46.7 million base pairs long. The human genome contains little over 3 billion base pairs that are constituents of 25,000 to 30,000 genes.

For anything to qualify as the genetic material, the essential characteristic is that it must be able to replicate true to type. How is DNA replicated? In their first paper on DNA structure, Watson and Crick wrote, "It has not escaped our notice that the specific pairing we have postulated, immediately suggests a possible copying mechanism for the genetic material". They elaborated in their second paper, "We imagine that prior to duplication the hydrogen bonds are broken, and the two DNA chains unwind and separate. Each chain then acts as a template for the formation on to itself of a new companion chain, so that eventually we shall have two pairs of chains, where we only had one before."

The implication is that each of the two double stranded DNA molecules that would be formed after replication of one double stranded DNA would have one old chain and one newly synthesized chain. Let us say the original double stranded DNA had two chains X-old / Y-old, where X-old and Y-old were complementary to each other according to the Watson and Crick base pairing rule, G across C and A across T. After replication, two double stranded DNA would be formed, X-old / Y-new and X-new / Y-old, where X-new and Y-new are the two new DNA chains, which have been synthesized following the Watson and Crick base pairing rule using X **–old** and Y **–old** as templates (Figure 26).. The base sequences of X-old and X-new are exactly the same, but X-old is the original DNA chain, while X-new is the newly synthesized chain. Similarly the base sequences of Y-old and Y-new are exactly the same, but Y-old is the original DNA chain, while Y-new is the newly synthesized chain.

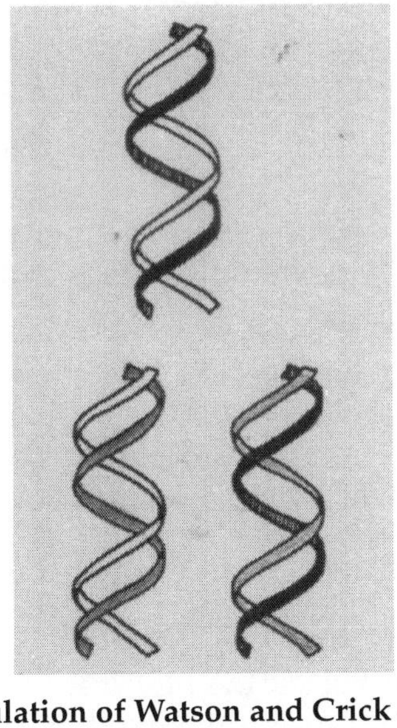

Legend of Figure 26. Postulation of Watson and Crick on replication of DNA. *Top double helix: Parent DNA, the black chain is chain* **X-old** *and the white chain is chain* **Y-old** *as mentioned in the text. Bottom two double helices: DNA after replication. Bottom left double helix: the white chain is chain* **Y-old** *and the dark grey chain is chain* **X-new.** *Bottom right double helix: the black chain is chain* **X-old** *and the light grey chain is chain* **Y-new.**

James Watson was offered a faculty position in Harvard University in 1956 and he continued there till 1976. From 1968 he had a concurrent appointment as Director of Cold Spring Harbor Laboratory. In 1994 he was made the President of the Cold Spring Harbor Laboratory.

Watson was awarded the John Collins Warren Prize of the Massachusetts General Hospital, the Lasker Award and the Nobel Prize. He is a member of the American Academy of Arts and Sciences, the National Academy of Sciences and the Danish Academy of Arts and Sciences.

In 2014, Watson startled the world by asking the auction house, Christie's to auction his Nobel prize medal. The Russian billionaire, Alisher Usmanov purchased it for U.S. $4.1 million, but later gifted the medal back to Watson.

Watson was raised as a Catholic, but as he had mentioned later was "an escapee from the Catholic religion". He had said, "The luckiest thing that ever happened to me was that my father didn't believe in God." Watson has strong, but controversial views about women in Science and human races. He wrote, "Anyone sincerely interested in understanding the imbalance in the representation of men and women in Science must reasonably be prepared at least to consider the extent to which nature may figure, even with the clear evidence that nurture is strongly implicated". About races, he had said, "stereotypes associated with racial and ethnic groups have a genetic basis: Jews being intelligent, Chinese being intelligent, but not creative because of selection for conformity". He had asserted that "differences in average measured IQ between blacks and whites are due to Genetics". James Watson was involved in a car accident in October 2018 and has not been keeping well since then.

Crick moved to the Salk Institute in La Jolla, California, U.S.A. in 1976 and worked on theoretical neurobiology and human consciousness. He remained there till his death in 2004 of colon cancer. In accordance with his wish, he was cremated and his ashes were scattered into the Pacific Ocean.

Crick was the recipient of the Nobel prize, the Lasker Foundation Award, the Royal and Copley medals of the Royal Society, the Benjamin Franklin Medal of the American Philosophical Society, the Prix Charles Leopold Meyer of the French Academy of Sciences, the Award of Merit of the Gairdner Foundation and Warren Triennial Prize. He was elected a Fellow of the Royal Society.

Annexure to Chapter 19.

Some details about the model of Watson and Crick

Watson and Crick constructed the physical scale model of DNA keeping in mind that (i) the bond angles and the bond distances were always correct, (ii) there would be no bad Van der Waal's contact, (iii) hydrogen bonds were allowed wherever possible and (iv) the physical scale model was space filling.

Foot note: Some details about DNA replication can be seen in Annexure of Chapter 19

The physical scale model had "two helical chains each coiled round the same axis".

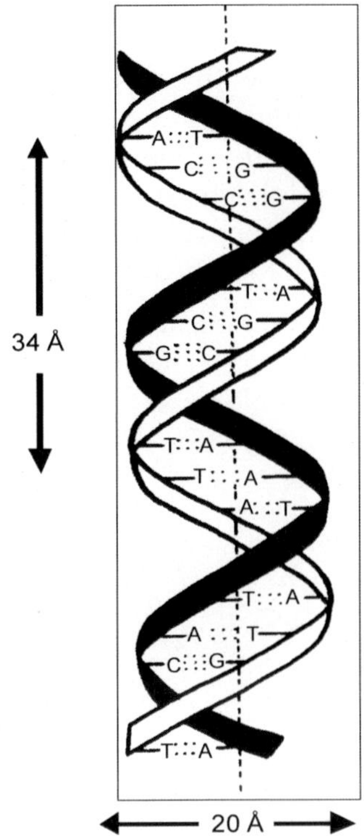

Legend of Figure 27. The double helix. *The helical ribbons represent the two phosphate – sugar chains, the horizontal pairs of bases holding the chains together. The figure is a reproduction of figure 52.5 from Chapter 52 entitled "The Structure of the Genetic Material and its Replication" of the book, "Textbook of Biochemistry and Human Biology" (ISBN:81-203-1965-6). Chapter 52 was contributed by H. K. Das. The book was edited by G. P. Talwar and L. M. Srivastava and published by Prentice-Hall India in 2002.*

"Each chain consists of phosphate diester groups joining β-D-deoxyribofuranose residues with 3′,5′ linkages. The two chains (but not their bases) are related by a dyad perpendicular to the fibre axis. Both chains follow right-handed helices, but owing to the dyad the sequences of the atoms in the two chains run in opposite directions". "There is a residue on each chain every 3.4 Å, in the *z*-direction. An angle of 36° has been assumed between adjacent residues in the same chain, so that the structure repeats after 10 residues on each chain, that is, after 34 Å. The distance of a phosphorus atom from the fibre axis is 10 Å." "The planes of the bases are perpendicular to the fibre axis. They are joined together in pairs, a single base from one chain being hydrogen-bonded to a single base from the other chain, so that the two lie side by side with identical *z*-coordinates. One of the pair must be a purine and the other a pyrimidine for [hydrogen] bonding to occur. The hydrogen bonds are made as follows: purine position 1 to pyrimidine position 1; purine position 6 to pyrimidine position 6". "The bases only occur in the structure in the most plausible tautomeric forms (that is, with the keto rather than the enol configurations). "Only specific pairs of bases can bond together. These pairs are: adenine (purine) with thymine (pyrimidine), and guanine (purine) with cytosine (pyrimidine)". Three hydrogen bonds are possible between guanine and cytosine, while two hydrogen bonds are possible between adenine and thymine. Watson and Crick pointed out that "the sequence of bases on a single chain does not appear to be restricted in any way". However, "if the sequence of bases on one chain is given, then the sequence on the other chain is automatically determined".

The following figure 28 is a non-helical representation of a hydrogen-bonded tetranucleotide duplex.

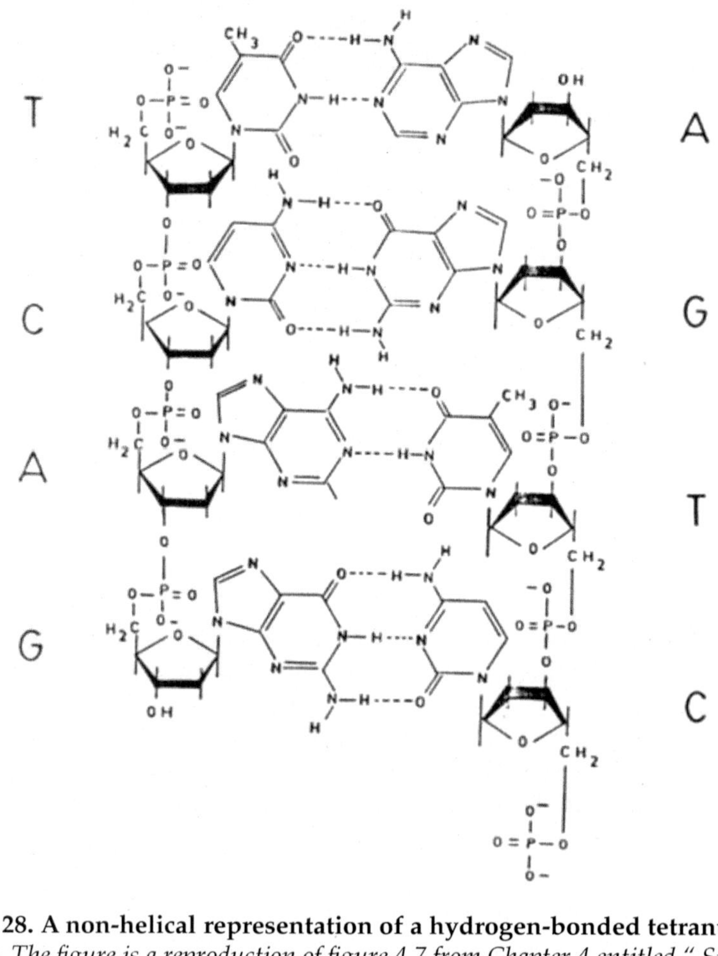

Figure 28. A non-helical representation of a hydrogen-bonded tetranucleotide duplex. *The figure is a reproduction of figure 4.7 from Chapter 4 entitled " Structure of DNA" of the book, "Gene and its Engineering" (ISBN: 978-81-265-4928-3). Chapter 4 was contributed by H. K. Das. The book was edited by H. K. Das and published by Wiley India Pvt. Ltd. in 2014.*

Some details about DNA replication

How is DNA replicated? In their first paper on DNA structure, Watson and Crick wrote, "It has not escaped our notice that the specific pairing we have postulated, immediately suggests a possible copying mechanism for the genetic material". They elaborated in their second paper, "We imagine that prior to duplication the hydrogen bonds are broken, and the two DNA chains unwind and separate. Each chain then acts as a template for the formation on to itself of a new companion chain, so that eventually we shall have two pairs of chains, where we only had one before."

What has been established later is that the two chains do not completely unwind at one go. DNA replication initiates by localized unwinding and formation of the "replication bubble" at a specific point(s) known as the replication origin(s). The replication bubble actually enlarges in both directions, finally resulting in the formation of two DNA double strands from one DNA double strand (Figure 29).

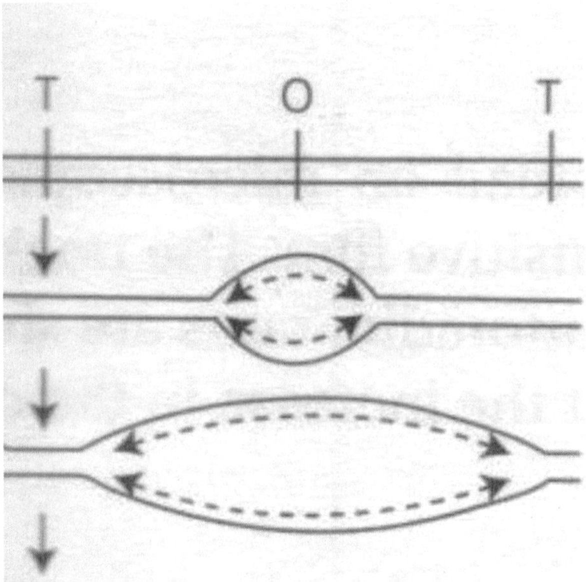

Legend of Figure 29. Replication bubble. *The O in top double strand represents the origin of replication. The middle double strand shows the initiation of the replication bubble, while the bottom double strand shows the enlarged replication bubble. The broken lines represent the newly synthesized DNA strands.*

The Figure 30 is a photograph that shows an actual DNA fork viewed through electron microscope. A replication fork is really a torn replication bubble.

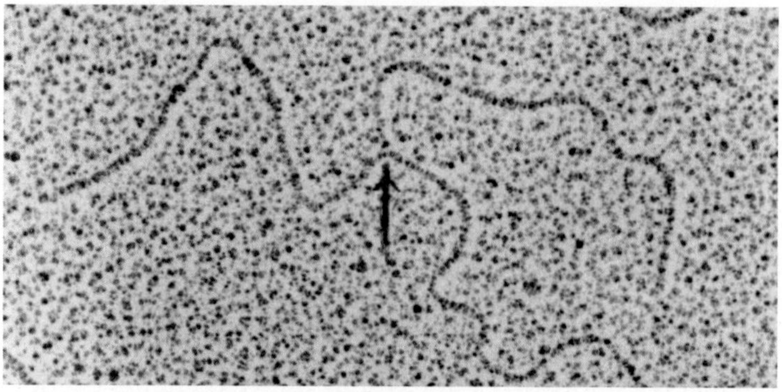

Legend of Figure 30. Photograph through electron microscope of an actual DNA fork, which is really a part of a torn replication bubble. *The DNA was stained with uranyl acetate, the position of the fork has been shown by the arrow. Magnification 1,00,000 fold. The photograph was taken in the laboratory of Prof. H. K. Das by Dr. Nirupam Roy Choudhury.*

In very simple terms, DNA is replicated using the existing two chains as templates and adding to the new strands deoxyribonucleotides one by one following the Watson and Crick base pairing rule, T, A , C, G in the new strand against A,T, G, C in the old strand, respectively, thus initiating formation of two new DNA chains. In reality, replication of one strand is discontinuous and lags behind the replication of the other strand, but for general understanding these are details that we do not have to worry.

20

FRANCIS CRICK, SYDNEY BRENNER AND CO-WORKERS DEMONSTRATED THAT THREE BASES IN DNA CODED FOR ONE AMINO ACID

Let us go back in time to 1952. A post-doctoral fellow named Sydney Brenner was working in the laboratory of Professor C.N. Hinshelwood in Oxford University. As Brenner wrote later in his autobiography, " I can remember in November 1952 Jack (Dunitz) telling me about two fellows in Cambridge who were going to solve the structure of DNA. When in April 1953, Jack told us that these two fellows in Cambridge, Francis Crick and Jim Watson, had indeed solved the structure of DNA, Jack, Leslie (Orgel) and I drove to Cambridge on a day between the 16th and 18th April to see the model. This was the watershed in my scientific life. The moment I saw the model and heard about the base pairs A across T and G across C, I realized that it was the key to understanding all the problems in biology we had found intractable – it was the birth of molecular biology. It was a revelation reinforced by conversing with Jim Watson at greater length during a walk we took together, when I realized that working with bacteriophage (viruses of bacteria) had put me on the right road to enter this exciting new field, even though what I was doing was trivial". Brenner also had a long discussion with Francis Crick, and both had planned to work together in future.

Brenner had come to Oxford from South Africa on a scholarship by the Commission for the Royal Exhibition of 1851. He returned to South Africa by the end of 1954, as that was the condition of his scholarship.

Sydney Brenner was born in 1927 in Germiston, a small town in South Africa. His parents were Jewish immigrants from Eastern Europe. His father was a shoe repairer, who did not know how to read or write. Sydney's father was persuaded by a customer, Miss Walkinshaw, to allow Sydney to join her kindergarten at the age of five free of charge. After completing kindergarten and primary school, Sydney went to the Germiston High School and matriculated in 1941 at the age of 14, two years earlier than his contemporaries. The Town Council of Germiston gave him a bursary of 60 pounds per year, which enabled him to join the University of the Witwatersrand in Johannesburg to study medicine.

Every morning he would cycle to the Germiston railway station, take a train to Johannesburg and walk to the University. The same way he would travel back home every evening. Around midway in his six year medical course, Sydney was informed that he was too young to qualify for medical practice after he completes the medical curriculum. So he deviated and earned a B.Sc. honours degree and then an M.Sc. degree, working part-time as a laboratory research assistant for financial support. It was during this period that Sydney realized that he enjoyed doing research much more than his medical curriculum. But after he obtained his medical degree, his parents wanted him to start medical practice and settle down. He, however, was keen to go back to England and do research, even though he did not have an easy life in England when he there last time. As Sydney Brenner had written in his autobiography, "There was still food rationing in England and life was difficult all through my 2 year stay in Oxford. In addition, I and the others were outsiders three times over; we were scientists, we were research students and we were colonials. Many of my friends in Oxford shared these stigmata and the only compensation was the opportunity to join Halifax House and lunch there." He still preferred research to medical practice. Around two years later Crick was able to get him an appointment in the Medical Research Council Unit in Cambridge. Brenner left for England again in December 1956 and teamed up with Crick there.

By then there was no doubt in any body's mind that DNA guided the synthesis of proteins, but there were only four bases in DNA.

Sydney Brenner

If each single base of the 4 different bases of DNA would code for one single amino acid, only $4^1 = 4$ amino acids would be coded. If any two bases would code for one amino acid, then $4^2 = 16$ amino acids would be coded. This is because the codes would then be AT, TA, GC, CG, AG, GA, AC, CA, TG, GT, TC, CT, AA, TT, GG and CC. There are, however, 20 different amino acids in a protein. Therefore, a minimum of three bases of DNA would be needed to code for an amino acid ($4^3 = 64$). That three bases of DNA indeed coded for an amino acid, was verified by an ingenious experiment done by Francis Crick, Sydney Brenner and co-workers.

The experiment and the inference of Crick, Brenner and co-workers were based on some facts and some assumptions.

The fact was that the language of DNA has only 4 letters, which are adenine (A), guanine (G), cytosine (C) and thymine (T). All words are composed of only these 4 letters. The assumptions were:

1. The language of DNA neither has any comma, nor does it have any space between the words.
2. All words have a fixed number of letters.
3. The words are read sequentially from a fixed point . If you assume that all words have three letters, you have to recognize the words by counting and reading three letters at a time one after the other.

During counting and reading if you miss one letter, all the subsequent words would be different or meaningless. For example, the sequence of three letter words, cat, rat, car, art and tar would appear in this new language as CATRATCARARTTAR, where the English letters C, A, T and R are equivalent to the 4 DNA bases, A, T, G and C of a gene.

If you start at the fixed point of the left most letter C and count 3 every time then you have no difficulty reading CAT RAT CAR ART TAR. Let us assume these five three letter words, which are part of a gene, code for the amino acids, U, V, W, Y and Z of the protein that is coded for by the gene.

If, however, you miss the A in CAT and then count 3 every time, you would read CTR ATC ARA RTT, but these words do not ring the bell, as these are very different. These new three letter words can not code for the original amino acids U, V, W, Y and Z, but let us say would code for the amino acids H, I, J and K.

Next if you miss the A of CAT and also miss the R of RAT, then you would read CTA TCA RAR TTA. These words again are very different. Let us say these three letter words would code for the amino acids P, Q, R and S.

Finally, if you also miss the T of CAT, you would miss three letters consecutively and you would read the meaningful words CAT CAR ART TAR. These would obviously code for the amino acids, U, W, Y and Z.

An important point to remember is that the activity of a protein is dependent on the amino acid sequence of the protein. If the sequence of amino acids of the protein is altered, the protein loses its activity. In the example mentioned above, you have restored the frame of 3 letter words by successively missing 3 letters and you have regained all the original amino acids of the protein. Even though you have lost just one amino acid V, chances are good that the protein would regain its activity, because all its other amino acids are unchanged.

The experimental material used by Crick, Brenner and co-workers was the virus T4. Viruses are very small and can not be seen even under the light microscope. One needs an electron microscope to visualize a virus . A virus comprises either DNA or RNA, wrapped up in a protein coat. Unlike a bacterial cell, a virus can not replicate by itself. For its replication, a virus must infect its specific host, which can be a plant or animal or a bacteria. The virus T4 has a double stranded DNA as its genetic material and its host is *Escherichia coli*, the bacteria present in our intestine.

One treatment of the virus T4 with the chemical proflavine was known to delete just one base pair from its DNA. Crick, Brenner and co-workers treated the virus T4 just once with proflavine and then infected *Escherichia coli* with the proflavine treated virus... They then determined the activity of a particular protein which the virus was supposed to produce after infection of *Escherichia coli*. They failed to detect any activity. Then they took the virus and treated it once again with proflavine and then infected *Escherichia coli* with the virus twice treated with proflavine. Still no activity of the protein. Next, they treated the virus the third time with proflavine and then infected *Escherichia coli* with the virus thrice treated with proflavine. This time the activity of the protein was restored. So the inference was that three bases coded for one amino acid and three successive one base pair deletions have restored the three base frame.

Sydney Brenner continued working in the Medical Research Council Unit in Cambridge and ultimately became the Director of the Unit in 1979 after the retirement of Max Perutz. Sydney worked there till 1992 and developed a new experimental system in the nervous system of the nematode, *Caenorhabditis elegans*. He then moved to the Scripps Research Institute in La Jolla, California. He was awarded the Nobel prize in 2002.

Brenner wrote in his autobiography, "I am still, at the age of 76, excited by scientific research and the prospect of what can be done in Biology. Science is something one is tied to for life and one should never retire from anything until one has secured one's next job. The endless quest for knowledge will continue as long as humans exist".

"Sydney was an atheist – as a child he was beaten up for being Jewish: 'I sat dazed and whispered a Hebrew prayer. But God didn't come. So that was the end of my relationship with him!' " Sydney Brenner died in 2019 at the age of 92.

Annexure to Chapter 20

Let us try to visualize how Crick and Brenner drew the inference from their experiment? A simple way would be to assume that the genetic code is indeed triplet and the actual codes are also known to us. Then go through the exercise and check whether the result matches.

An important point to remember is that the activity of a an enzymic protein is dependent on the amino acid sequence of the protein. If the sequence of amino acids of the protein is altered, the protein loses its activity. Let us now check whether three successive deletions of one base pair of DNA really restores the amino acid sequence of the protein, thus helping it regain its activity.

Foot note: Annexure to Chapter 20 contains an exercise assuming that the genetic code is indeed triplet and then check if the result of three successive single base deletion matches.

The result of such an exercise is elaborated in Figure 31. Though Crick, Brenner and other co-workers had drawn their inference without any idea as to which amino acid was coded for by which triplet of bases, for our easy understanding of their experiment, we have assumed that we have full knowledge of the codes.

Stage 1

↑

ATG GCT CAG CCA TGC ACG CAT GTG GAA TCG ATC TTC TGG CTG GAC CGT GGT CAC TAC
Met Ala Gln Pro Cys Thr His Val Glu Ser Ile Phe Trp Leu Asp Arg Gly His Tyr

↓

Stage 2 ↓

↑ ↓

ATG CTC AGC CAT GCA CGC ATG TGG AAT CGA TCT TCT GGC TGG ACC GTG GTC ACT AC
Met Leu Ser His Ala Arg Met Trp Asn Arg Ser Ser Gly Trp Thr Val Val Thr

↓

Stage 3 ↓

↑ ↓

ATG CTC AGC ATG CAC GCA TGT GGA ATC GAT CTT CTG GCT GGA CCG TGG TCA CTA C
Met Leu Ser Met His Ala Cys Gly Ileu Asp Leu Leu Ala Gly Pro Trp Ser Leu

↓

↓

Stage 4 ↓

ATG CTC AGC AGC ACG CAT GTG GAA TCG ATC TTC TGG CTG GAC CGT GGT CAC TAC
Met Leu Ser Ser Thr His Val Glu Ser Ile Phe Trp Leu Asp Arg Gly His Tyr

Legend of Figure 31. A simplistic representation of the Crick-Brenner experiment suggesting triplet code. *The top line in stage 1 is the representation of the base sequence of a part of the coding strand of the DNA of the virus T4. Assuming triplet code, every three consecutive bases have been grouped together and coloured differently. The amino acid coded for is written below each triplet of bases in the bottom line of stage 1. Following are the abbreviations of the names of the amino acids: Met, methionine; Ala, alanine; Gln, glutamine; Pro, proline; Cys, cysteine; Thr, threonine; His, histidine; Val, valine; Glu, glutamic acid; Ser, serine; Ile, isoleucine; Phe, phenylalanine; Trp, tryptophan; Leu, leucine; Asp, aspartic acid; Arg, arginine; Gly, glycine; His, histidine; Tyr, tyrosine. The green arrow pointing upwards above the base G in stage 1 indicates that this base would be deleted. The top line in stage 2 represents the sequence of bases after the deletion of the base G. Every three consecutive bases have been grouped together again assuming triplet code. It is obvious that the groupings are altered after deletion of the base G in stage 1 and hence the composition of the triplets are also* *altered. Consequently, the amino acids coded for by the triplets (assuming triplet code), as shown in the bottom line of stage 2, are also altered. As a result the protein loses its activity. The blue arrow pointing upwards above the base C of the top line in stage 2 indicates that this base would be deleted next. The top line in stage 3 represents the sequence of bases after the deletion of the base C in stage 2. Here again every three consecutive bases have been grouped together. It is obvious that the groupings are altered once again after the deleted base and hence the composition of the triplets are also altered once again. Consequently, the amino acids coded for by the triplets, as shown in the bottom line of stage 3, are also altered once more. As the altered amino acid sequence is still not the same as the original sequence of amino acids, the protein still remains inactive. The yellow arrow upwards above the base T of the top line in stage 3 indicates that this base would be deleted next. The top line in stage 4 represents the sequence of bases after the deletion of the base T in stage 3. Once again every three consecutive bases have been grouped together. It is obvious that the groupings are altered after the deleted base and hence the composition of the triplets are also altered once more. Consequently, the amino acids coded for by the triplets, as shown in the bottom line of stage 4, are also altered and are different from the sequence of amino acids in stage 3.. It is interesting, however, to note that the sequence of amino acids in stage 4 beginning the 5th amino acid from the left is exactly the same as the sequence of amino acids in stage 1 beginning the 6th amino acid from the left.*

The experiment described in Figure 31 thus shows that after three consecutive single base deletions, the sequence of amino acids of the protein, is almost completely restored to the original. Hence the activity of the protein is also restored after deletion of three bases. It thus implies that the amino acid code indeed comprises three bases of the DNA, or the language of DNA comprises only 3-letter words. It has to be appreciated that Crick, Brenner and co-workers drew the inference of the code being triplet, just by logical analysis of their result, without having any idea about the actual codes..

21

FRANCIS CRICK'S CENTRAL DOGMA OF MOLECULAR BIOLOGY

The work of Oswald Avery had established that the gene was nothing but DNA. George Beadle and Edward Tatum demonstrated that enzymes were being synthesized under the guidance of genes. James Sumner revealed that enzymes comprised proteins. Thanks to the work of Jerardus Mulder, Franz Hofmeister, Emil Fischer, Erwin Chargaff, Rosalind Franklin, Francis Crick, James Watson and Sydney Brenner, we know quite a bit about proteins and DNA. The next question then is whether DNA directly guides the synthesis of protein? The answer to this question has a bearing on where the DNA resides in a cell. Figure 32 shows an artist's simplistic impression of an animal cell viewed through a microscope, showing the nucleus surrounded by the cytoplasm.

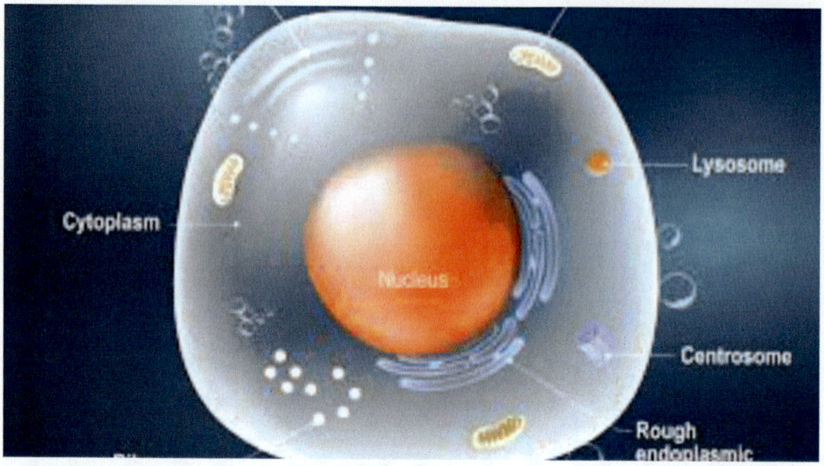

Legend of Figure 32. An artist's simplistic impression of an animal cell viewed through a microscope, showing the nucleus surrounded by the cytoplasm. The colours are not real.

Using specific stains, most of the DNA has been found located inside the nucleus, which is surrounded by the nuclear membrane. Protein biosynthesis, however, takes place in the cytoplasm of the cell, outside the nuclear membrane. How is the message from the DNA inside the nucleus communicated to the protein synthesizing machinery in the cytoplasm outside the nucleus? May be, there is a group of molecules that originate inside the nucleus and has the information of base sequence of DNA, which travel through the pores in the nuclear membrane into the cytoplasm and direct the amino acid sequence of the proteins at the site of protein synthesis. The simplest thing to imagine for us would be that the said group of molecules would be similar to DNA and would have the imprint of the base sequence of DNA. But at that point of time, nobody had an exact idea of how the communication from DNA to protein actually occurred.

It was Francis Crick again who came up with a hypothesis. Crick enunciated the Central Dogma of Molecular Biology in his lecture at the 1957 annual meeting of the Society of Experimental Biology held in London. In lay man's term it said, DNA → RNA → protein. More precisely, it stated that genetic information stored in DNA can flow into protein through RNA, but it can not flow directly from DNA to protein. As before, he did not do any laboratory experiment for developing this idea, it was just making use of his razor sharp intellect.

22

WHAT IS RNA?

Indeed in all cells there is a group of molecules similar to DNA which is called RNA (ribonucleic acid). The structure of RNA is similar to that of DNA. RNA is also a long linear molecule, which comprises a chain of ribose – phosphate – ribose – phosphate – ribose – phosphate – and so on with the purine or pyrimidine base connected at the other end of each ribose moiety. So in RNA the sugar is ribose instead of deoxyribose, which is present in DNA. The purine base of RNA is also either adenine (A) or guanine (G), but the pyrimidine base is either uracil (U) or cytosine (C). So one of the pyrimidines is uracil, instead of thymine, which is present in DNA. Similar to the snake analogy of the DNA chain (chapter 16), we can also compare the RNA chain with a snake.

The RNA snake has alternate black and grey bands on its body and on its back on every grey band, it carries a load. The head of the snake symbolises the 5' end of the RNA. Immediately behind its head the snake has a black band which represents the phosphate moiety of RNA. The next band is grey which represents the ribose (R) moiety of RNA (instead of deoxyribose in DNA). The load on the grey band can be the purine adenine (A) or guanine (G), or the pyrimidine cytosine (C) or uracil (U). Next to the grey band there is a black band, then again a grey band, then again a black band, then again a grey band and so on. Finally the tail of the snake is there which symbolises the 3' end of RNA. The snake is lying on its belly (Figure. 33)..

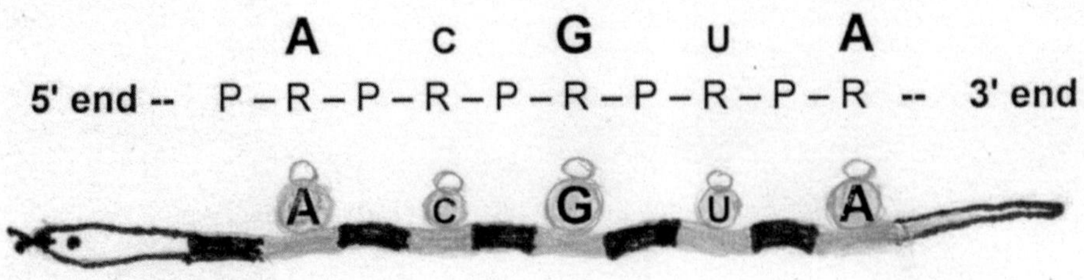

Legend of Figure 33. Analogy of a segment of RNA (top element) with a snake (bottom element)..

Thus, there are two differences between DNA and RNA. One is the sugar, it is deoxyribose in DNA, while it is ribose in RNA. The other difference is in one of the pyrimidine bases, it is thymine in DNA, while it is uracil in RNA. Structurally uracil and thymine are very similar. RNA in cells, however, is always single stranded. As has been mentioned in chapter 19, DNA in cells is always double stranded.

23

OBSERVATIONS OF BRACHET IMPLICATED INVOLVEMENT OF RNA IN PROTEIN SYNTHESIS

Francis Crick hypothesized in 1957 in London that RNA .serves as an intermediate in the process of DNA guided protein synthesis. As we were busy admiring Crick, it dawned on us that he did not derive his hypothesis out of vacuum. More than two decades earlier, across the English channel, in the city of Brussels Jean Louis Brachet, Professor of Animal Morphology and General Biology at the Université Libre de Bruxelles, Belgium, had obtained indication that DNA did not directly guide protein synthesis, but probably did it through RNA. Brachet, who was looking at cells of yeasts and various plants and animals through his microscope, had developed a staining protocol by which he could colour RNA and DNA in a cell differently. Using this protocol he had established in 1933 that both DNA and RNA were present in all cells, irrespective of its source, but while DNA was mostly in the nucleus, RNA was in the cytoplasm. Brachet also observed that while DNA levels were more or less unchanged, RNA levels increased in the cells that were actively synthesizing proteins, thus implicating involvement of RNA in protein synthesis.

Jean Louis Brachet was born in 1909 in Etterbeek near Brussels in Belgium. He was the son of Albert Brachet. Albert was teaching Anatomy and Embryology at the School of Medicine in Paris in 1914, when the World War I broke out and the German army invaded France. The Brachets were stranded in France. After the war, they moved back to Belgium and Jean graduated in medicine 1934 from the Universite Libre de Bruxelles. Very much like his father, Jean Brachet joined the teaching profession, rather than starting medical practice after his graduation, and also started scientific research. Soon he rose to become Professor of Animal Morphology and General Biology at the Université Libre de Bruxelles.

Jean Louis Brachet

Brachet had also found that a specific RNA fraction first appeared in the nucleus and then got transferred into the cytoplasm, suggesting that this RNA fraction was the carrier of genetic information. When they treated the cells with the enzyme ribonuclease, which degrades RNA, protein synthesis was inhibited. Finally, they removed the nucleus of the alga *Acetabularia*, thus eliminating its DNA, but found that protein synthesis continued in the cytoplasm of the cell for some days before ultimately stopping.. Brachet may not have realized the implication of his findings, but this was the decisive indication that DNA probably guided protein synthesis through a species of RNA that was not very stable.

World War II broke out in 1939 and Belgium was occupied by the German army. The Université Libre de Bruxelles was closed down. Jean Brachet was arrested and all his research work came to a stop. He resumed his research work as Belgium was liberated at the end of the war in 1943. Brachet then became the Research Director of the International Laboratory for Genetics and Biophysics in Naples, Italy.. He retired in 1977, but continued his research activity splitting his time between Brussels and Naples. Jean Brachet was awarded the Francqui Prize for Biological and Medical Sciences. He passed away in 1988.

24

FRANÇOIS JACOB AND JACQUES MONOD MOOTED THE CONCEPT OF MESSENGER RNA AS CARRIER OF MESSAGE FROM DNA

Across the border of Belgium, in the city of Paris in France, Francois Jacob was studying medicine at the Université de Paris Pierre et Marie Curie after finishing school in 1937. He wanted to be a surgeon mainly because of his admiration of his uncle, who was a fine surgeon. Francois Jacob was born in a Jewish family in 1920 in the small town Nancy in France. His father was a businessman and his mother was the daughter of a decorated General. Francois described himself "a shade backwards with girls." With boys, the right-wing bullies were often after him because of his Jewish decent.

France was also over run by the German army during the World War II. Francois could not accept this humiliation of defeat to the Germans. In 1940 he temporarily abandoned his dream of becoming a surgeon and quietly slipped out of France to join the Free France Forces in England organized by Charles de Gaulle with the aim of liberating France. For four years he worked with the Allied Forces that were fighting the German army under Rommel in the African countries of Chand, Libya and Tunisia. In August 1944 he was with the Allied Liberation Forces that landed on Normandy beach of France, but was severely wounded in the operation.

His recovery took long seven months. After the liberation of France, he was awarded Croix de la Liberation, the highest military award. He resumed his medical studies and submitted the doctoral thesis in medicine in 1947. He could not become a surgeon because of his injuries and he did not want to be a medical practitioner. Jacob, as he described in his autobiography, was finding it difficult to return to a "normal life". He considered various possibilities, be an administrator, enter politics, or become an actor. Finally, he joined a project aimed at producing antibiotics, but the project was a failure. A cousin, who worked in a laboratory, suggested Jacob to try his hand in Biology research. In 1950 at the age of 30 he secured a research position in the laboratory of Andre Lwoff, who worked in an attic in a very old building of the Institut Pasteur in Paris. There Jacob worked for a doctorate degree in Science and obtained it in 1954. He became the laboratory Director in 1956. In 1957 he initiated his collaboration with Jacques Monod, who was working in the laboratory down the hall.

Francois Jacob

Jacob's collaborator, Jacques Lucien Monod was born in Paris in1910 in a family of professors, civil servants, pastors, and doctors. His father, however, was a painter and art historian. His mother was the daughter of Scottish pastor who had migrated to the U.S.A. Jacques grew up in an environment that was intellectual, artistic, and musical. He liked playing cello, climbing rocks and sailing yachts. After secondary education, he went to Sorbonne, the famous University in Paris, and obtained the bachelor's degree in Science in 1931. While in Sorbonne, he founded a Bach choral group, "La cantateI" and toyed with the idea of becoming a music conductor. Meanwhile he obtained a research grant and his career in Biology research was initiated. He participated in several research projects, but a career in research did not really enamour Jacques Monod. He decided to join Commander Charcot and go off to a scientific expedition in Greenland in 1936 to study its natural history on the boat, "*Pourquoi pas*". At that point of time, one of his research supervisors, Boris Ephrussi, was about to leave for a year to work in the California Institute of Technology in Pasadena, U.S.A.. Ephrussi dissuaded Monod from leaving research and arranged for a Rockefeller Foundation grant for him, so that Monod could accompany Ephrussi and learn Genetics in Pasadena. This saved Monod's life, as *Pourquoi pas* sank while negotiating a severe storm off the coast of Greenland. Monod, however, was not keen to learn Genetics, but got involved in participating in orchestras and choral groups in Pasadena. Returning to Paris after a year, he resumed playing the cello in a quartet and directing the Bach choral group, while working as a Research Assistant in Sorbonne. Nevertheless, he eventually started finding interest in research and in 1941 obtained his Science doctorate.

Jacques Lucien Monod was working as a Research Assistant in Sorbonne, the famous University in Paris, when the Germans occupied France during World War II. Monod interrupted his research carrier and joined the underground resistance movement against the German occupation army. In 1943 he became the Chief of the National Staff of Paris. The Germans had captured and executed three of his predecessor Chiefs. Monod evaded the Germans and mobilized the people in Paris to mount barricades and facilitate liberation of Paris by the allied forces. After liberation of France, Monod returned to his laboratory in André Lwoff's department at the Institut Pasteur.

Jacques Lucien Monod

François Jacob and Jacques Monod experimented in their laboratory with *Escherichia coli*, the friendly bacteria that all of us have in our gut. The bacteria *Escherichia coli* can be easily grown in the laboratory in a medium containing the simple sugar glucose, common salt and a few other minerals. In this medium *Escherichia coli* divides every 40 to 50 minutes at 37°C. The protein they studied synthesis of in *Escherichia coli*, was the enzyme β-galactosidase, which splits the sugar lactose. This enzyme was found by them to be present in *Escherichia coli* in a very low amount, when the bacteria were grown in a medium containing the sugar glucose. The bacteria, however, synthesized a large amount of β-galactosidase, when glucose was replaced by the sugar lactose in the growth medium.

Making a long story short, they established that the gene that coded for the enzyme β-galactosidase remained repressed or shut down by the action of an agent they termed repressor, when the bacteria were grown in glucose. When glucose was replaced by lactose, it combined with and removed the repressor and thus freed the gene to express or function, enabling the bacteria to synthesize the enzyme β-galactosidase.

Next, they used an uracil mutant of *Escherichia coli*, a strain of *Escherichia coli* that could not biosynthesize uracil. The medium of growth of this mutant had to be always supplemented with uracil. This mutant also synthesized a large amount of β-galactosidase, when both uracil and lactose were present in the growth medium. If, however, the mutant *Escherichia coli* cells were washed and resuspended in a medium containing lactose but no uracil, synthesis of the enzyme β-galactosidase stopped. The synthesis of β-galactosidase could be resumed if uracil was added back.

So presence of uracil was essential for the synthesis of β-galactosidase. But β-galactosidase is an enzyme, which is a protein and contains only amino acids. It does not contain any uracil. As has been described in chapter 21, uracil is a constituent of RNA. Jacob and Monod hypothesized that synthesis of a specific type of RNA was essential for the synthesis of the enzyme β-galactosidase. The named this type of RNA as messenger RNA, implying that this type of RNA carries the message from DNA for the synthesis of the protein. The messenger RNA is not stable unlike other types of RNA of the cell, that is why it has to be continuously synthesized for the synthesis of the protein.

Francois Jacob and Jacques Monod were awarded the Nobel prize in 1965 along with Andre Lwoff. In his Nobel lecture Jacob said, "Every one of us dreamed of changing the world at twenty. Every one of us knows that s(h)e won't do it at the age of forty. At best and with a little luck we can hope to contribute in some way to the stock of truths, recipes and ideas that human beings have been slowly but surely accumulating over time". Jacob was a foreign member of the American Academy of Arts and Science and the American Philosophical Society. He was awarded a seat in the Academie Francaise in 1996. Francois Jacob died in 2013 at the age of 93.

Monod was made head of a new department called Cellular Biochemistry in 1953. Monod became Director of the Pasteur Institute in 1971.. In addition to the Nobel prize in 1965, Monod was awarded the Louis Rapkine Medal in London in 1958, the Legion d'Honneur, one of the highest distinctions possible for a Frenchman, in 1963.. He was elected a Foreign Member of the American Academy of Arts and Sciences in 1960 and of the Royal Society in 1968. Monod died of leukemia in 1976 at the age of 66.

25

SEARCH FOR THE HYPOTHETICAL MESSENGER RNA OF JACOB AND MONOD

It was April 15 1960, Good Friday, from France we had crossed the English channel back to England and drove to Cambridge. It was early evening. We peeped into the house of Francis Crick. He and his wife had thrown a party.. François Jacob was also there, he had come to London from Paris to attend a conference that was held the previous day. Jacob sat in one corner with Sydney Brenner discussing intensely a plan to lay their hands on messenger RNA that he and Monod had hypothesized, trying hard not to be distracted. As Jacob recalled, "A very British evening with the cream of Cambridge, an abundance of pretty girls, various kinds of drink, and pop music. Sydney and I, however, were much too busy and excited to take an active part in the festivities…It was difficult to isolate ourselves at such a brilliant, lively gathering, with all the people crowding around us, talking, shouting, laughing, singing, dancing. Nevertheless, squeezed up next to a little table as though on a desert island, we went on, in the rhythm of our own excitement, discussing our new model and the preparations for experiment…A euphoric Sydney covered entire pages with calculations and diagrams. Sometimes Francis would stick his head in for a moment to explain what we had to do. From time to time, one of us would go off for drinks and sandwiches. Then our duet took off again".

The crux of the problem was that a cell contains variety of RNAs, the messenger RNA would have to be distinguished from other RNAs present in an *Escherichia coli* cell. Let us try to understand what other RNAs are present in any cell. In all cells there are some particulate entities named ribosomes, which are complexes of a few RNAs and a large number of proteins. The ribosomes of *Escherichia coli* contain three RNA species, the largest one has about 3,000 bases, a smaller one that has about 1,500 bases, and a smallest one which has about 100 bases. Besides these, some soluble RNAs containing about 80 bases are also there in a cell of *Escherichia coli*. What would be the size of messenger RNA? The messenger RNA are supposed to code for all the proteins in the cell. Proteins are of various sizes, the average being about 300 amino acids. Sydney Brenner and Francis Crick had demonstrated that three bases in DNA coded for one amino acid (chapter 20). Since messenger RNA carries the code from DNA to RNA (chapter 24), three bases of messenger RNA would also code for one amino acid. Then the average size of messenger RNA would be expected to be about 900 bases. The messenger RNA, therefore, are lighter than some cellular RNA, but heavier than some other cellular RNA. The RNA molecules, however, are very small, simple gravity would not be of any use to separate these molecules. The equipment needed for separation of such small molecules is an ultra high speed centrifuge which works on the principle of sling shot. Higher the speed of rotation, greater is the force generated. The centrifuge rotor would need to operate at about 25,000 revolutions per minute for about 15 hours to separate the RNA molecules. Besides, special photograph taking device would have to be there to visualize the positions of the separated RNA molecules. At that point of time such an equipment was not available in England. Mathew Meselson had the required equipment in the California Institute of Technology, Pasadena, U.S.A.. Jacob and Brenner travelled to Pasadena as Meselson agreed to collaborate with them. After several months of intense work they were successful in isolating messenger RNA, an RNA species that appeared between the 1,500-base small ribosomal RNA and the 80-base soluble RNA, an RNA species that had been copied from the DNA. Mission accomplished, Jacob and Brenner went back to their own laboratories. Brenner completed writing the paper and was about to communicate it to the journal Nature.

Meanwhile a group of scientists in Harvard University in U.S.A. under the supervision of James Watson were also trying to get at messenger RNA and had almost achieved their objective. Somehow the news reached Watson that Brenner was about to communicate their results, so Watson had lost the race. He was crestfallen and desperate. He talked to Jacob and sent a telegram to Brenner requesting him to withhold communicating their paper, till the paper of Watson's group was ready. A copy of the telegram is shown in Figure 34. Jacob and Brenner were generous enough to agree and the two papers appeared back to back in May 1961.

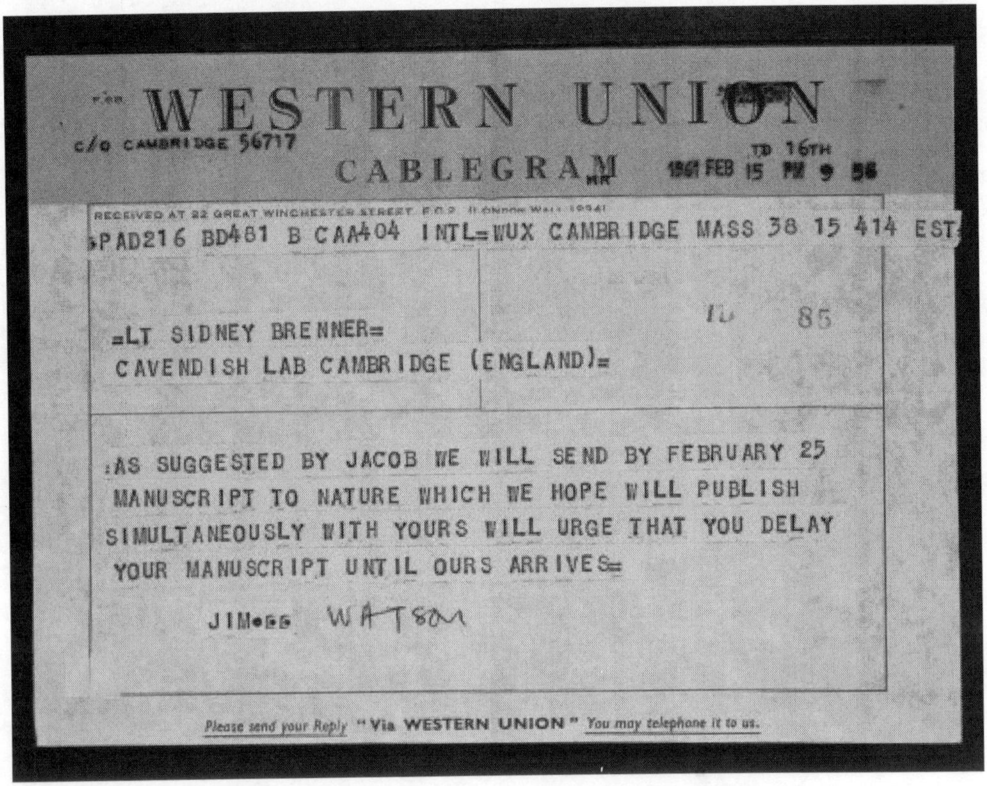

WESTERN UNION

C/O CAMBRIDGE 56717

CABLEGRAM 1961 FEB 15 PM 9 56

TD 16TH

RECEIVED AT 22 GREAT WINCHESTER STREET, E.C.2

PAD216 BD461 B CAA404 INTL=WUX CAMBRIDGE MASS 38 15 414 EST

=LT SIDNEY BRENNER=
CAVENDISH LAB CAMBRIDGE (ENGLAND)=

AS SUGGESTED BY JACOB WE WILL SEND BY FEBRUARY 25
MANUSCRIPT TO NATURE WHICH WE HOPE WILL PUBLISH
SIMULTANEOUSLY WITH YOURS WILL URGE THAT YOU DELAY
YOUR MANUSCRIPT UNTIL OURS ARRIVES=

JIM=55 WATSON

Please send your Reply "Via WESTERN UNION" You may telephone it to us.

Figure 34. Copy of the telegram that was sent by James Watson requesting Sydney Brenner to delay publication of their manuscript.

It may be mentioned at this point that of the two strands of DNA, one specific strand serves as the template for the messenger RNA to be synthesized. This strand is known as the coding strand of DNA. Watson and Crick's base pairing rule is also followed during the synthesis of messenger RNA. The DNA duplex shown in chapter 19 is being reproduced below.

Top strand T-C-T-G-A-C-T-C-A-G-T-G-A-C-A-T-G-T (non-coding strand)

Bottom strand A-G-A-C-T-G-A-G-T-C-A-C-T-G-T-A-C-A (coding strand)

If the bottom strand of the DNA is the coding strand, then the messenger RNA synthesized would have the following base sequence:

U-C-**U**-G-A-C-**U**-C-A-G-**U**-G-A-C-A-**U**-G-**U**

It may be seen that the base sequence of the messenger RNA is the same as the base sequence of the non-coding strand of the DNA, except that all the Ts (thymines) have been replaced by Us (uracils).

26

THE RNA TIE CLUB

The Central Dogma of Molecular Biology enunciated by Francis Crick stated that genetic information stored in DNA flows into protein through RNA. Francois Jacob and Jacques Monod hypothesized that synthesis of a specific type of RNA, which they named messenger RNA, was essential for the synthesis of proteins. Francois Jacob, Sydney Brenner and Mathew Meselson and also James Watson and co-workers demonstrated the actual existence of messenger RNA. So now all attention was on RNA. James Watson and Russian–American physicist George Gamow founded the RNA Tie Club to "solve the riddle of the RNA structure and to understand how it built proteins". There were 20 members of the Club corresponding to the 20 amino acids present in proteins. Each member received a black woolen necktie with a green and yellow RNA helix embroidered on it. Members of the club also received a golden tiepin with the three letter abbreviation of the member's amino acid designation. George Gamow was allotted the amino acid, alanine, so his tiepin had the letters ALA written on it. Erwin Chargaff's tiepin had the letters LYS written on it, as he was allotted the amino acid lysine. James Watson got the amino acid proline, so his tiepin had the letters PRO written on it. Francis Crick's tiepin had the letters TYR written on it, as he was allotted the amino acid tyrosine. Sydney Brenner was allotted the amino acid valine and his tiepin had the letters VAL written on it. Whenever any member would wear the Club's tiepin, people not aware of the Club would wonder why they were wearing a tiepin mentioning somebody else's initials. The meetings of the Tie Club were held twice a year, but between meetings members would exchange new concepts and ideas.

27

MARSHALL NIRENBERG INITIATED CRACKING OF THE GENETIC CODE AS A FIRST STEP IN UNDERSTANDING HOW PROTEINS ARE MADE

Sydney Brenner and co-workers established that the amino acid code is a triplet of bases. But how to know the triplet codes of all the amino acids?

It was August, 1961, we had travelled to Moscow. We found about one thousand Biochemists and Molecular Biologists from all over the world had assembled for attending the International Congress of Biochemistry. A young handsome gentleman was nervously pacing up and down outside a lecture room clutching the draft of his talk, as he was the next speaker. This was 34 year old Marshall Nirenberg, a stranger among the great Biochemists and Molecular Biologists. Nirenberg was not a member of the RNA Tie Club. As he said later, "I didn't know the people in Molecular Biology... I didn't know anybody in protein synthesis... I was working on my own."

Only 35 people attended Nirenberg's talk and they were "absolutely dead". One person, however, appreciated the importance of Nirenberg's results, and he reported it to James Watson. Watson told Francis Crick. Crick persuaded the organizers of the Congress to invite Nirenberg to repeat his talk to the general audience of about 1000 scientists attending the Congress. "The reaction was incredible," Nirenberg remembered. "It was a standing ovation... I didn't know it then, but for the next five years I became like a scientific rock star."

Marshall Warren Nirenberg was born in 1927 in New York in a Jewish family. His father was a shirt maker. As Marshall suffered from recurrent rheumatic fever in his boyhood, his father moved the family to a farm of warmer Orlando in Florida. Marshall was having fun there watching birds, digging snakes out of their burrows and collecting spiders.

After finishing school, Marshall went to the University of Florida at Gainesville. His thesis for the Masters degree was on caddis flies, a moth like flying insect found in the marshes of Florida.. Marshall moved to the University of Michigan at Ann Arbor for his doctoral work and earned his Ph.D. in Biochemistry in 1957. He then joined the National Institutes of Health as a Research Fellow.

Nirenberg and his post-doctoral fellow, J. Heinrich Matthaei broke open the cells of the colon bacteria *Escherichia coli* and used the lysate that was free of intact cells, as source of the protein synthesizing machinery. They used radioactively labelled amino acids to track their incorporation into proteins and thus measure protein synthesis.

Protein synthesis was rather low. Nirenberg suspected that it was due to inadequacy of messenger RNA in the cell lysate. When they added messenger RNA isolated from *Escherichia coli*, protein synthesis was rapid. When they added DNA from *Escherichia coli*, there was no improvement in protein synthesis. That answered the age old question once again, DNA did not directly guide protein synthesis, but it did it through messenger RNA.

Next, Nirenberg obtained polyuracil, the homo-polymer of uracil, which may be considered as an artificial messenger RNA containing only uracil instead of all the four bases, uracil (U), adenine (A), guanine (G), cytosine (C). He wrote up a detailed protocol for its use and in the evening requested Matthaei to conduct the experiment. Matthaei added the polyuracil to the cell lysate in 20 test tubes. Then he added to each tube a different amino acid that was radioactively labelled. "3:00 in the morning, May 27, 1961, a Saturday", Matthaei saw that the amino acid got incorporated into protein in only one tube, the one that received radioactive phenylalanine.

He analysed the radioactive protein and found it to be a polymer of phenylalanine. Thus polyuracil acted as the messenger RNA for polyphenyl alanine. Since the code was supposed to be a triplet, the code for phenylalanine must then be UUU. This was the first example of the genetic code unravelled. "That was really staggering," Nirenberg recalled later. Subsequently, Nirenberg obtained synthetic ribohomopolymers of adenine and cytosine and found out that they acted as messenger RNA for polylysine and polyproline respectively. Thus the codes for the amino acids lysine and proline were AAA and CCC respectively.

The crucial point was that the endogenous protein synthesis by the cell lysate of *Escherichia coli* was quite low, because of the absence of endogenous messenger RNA in the lysate. The reason was, as it was known later, messenger RNAs of *Escherichia coli* get degraded fast.

The tube containing the externally added artificial messenger RNA polyuracil, elicited huge amount of protein synthesis, while all the other tubes exhibited practically no protein synthesis. That is why the result was so convincing. If Nirenberg had used lysate of rat liver, instead of lysate of *Escherichia coli*, he would have seen protein synthesis in all the 20 tubes, because rat liver lysate contains enough of endogenous messenger RNA, as rat liver messenger RNA is not degraded that fast. The tube containing phenylalanine would then have elicited only some enhanced protein synthesis. The result would not have been "staggering".

Paul Zamecnik indeed observed enough endogenous protein synthesis by rat liver lysate, irrespective of what radioactive amino acid he used (Chapter 32). Endogenous proteins comprise all 20 amino acids. So the biological system that one chooses to use also determines the chances of success.

Nirenberg was then working in the National Institutes of Health in Bethesda, Maryland, U.S.A. By 1966, Nirenberg with the help of his new post-doctoral fellow, Phillip Leder, had found out all the three-base codons for each of the 20 amino acids present in proteins by synthesizing the codons biochemically using an enzyme-based approach.

"This was first glimpse of the biological Rosetta Stone," said NIH Director Francis Collins, commenting on Nirenberg's discovery that polyuracil coded for polyphenyl alanine. Nirenberg's work "sent shock waves through the scientific community". "He didn't stop with cracking the genetic code, but went on to prove it was universal." Marshall Nirenberg was a "scientist's scientist" and a "mentor's mentor".

Once he completed cracking the genetic code, Nirenberg changed fields and shifted to neurobiology. His work on the assembly of the nervous system in the fruit fly *Drosophila melanogaster* proved to be very relevant to studies on the development of the nervous system in humans.

Marshall Warren Nirenberg

In 1960 Nirenberg was promoted to the independent position of Research Biochemist. "Starting out as an independent investigator you are supposed to hit the deck running and prove that you are an effective, productive investigator," Nirenberg said later. "On the one hand I wanted to explore an important problem, on the other I was afraid of failure, but the wish to explore was much greater than the fear of failure."

"Marshall had the curiosity of a brilliant child," said Myrna Weissman who was a Professor of Epidemiology and Psychiatry at Columbia University College of Physicians and Surgeons in New York. He had "an incredible imagination [and] was a dreamer." She went on to extol his virtues, adding that he had a complete "lack of arrogance [and] was a most compassionate man."

Nirenberg was awarded the National Medal of Science, National Medal of Honour and the Nobel prize. He was elected to the American Philosophical Society . Nirenberg died of cancer in 2010 at the age of 82.

Annexure to Chapter 27

Following was a part of Nirenberg's draft of his talk delivered in the International Congress of Biochemistry held in August 1961:

"We have obtained a stable, cell-free *E. coli* system which incorporates C^{14} amino acids into protein at a rapid rate and which has many characteristics of protein synthesis. Conditions have been found which demonstrate a novel characteristic of this system, that is, a requirement for messenger RNA, needed even in the presence of excess soluble RNA and ribosomes. Naturally occurring RNA, such as *E. coli* messenger RNA or a synthetic polynucleotide, were active in this system. The synthetic polynucleotide seems to contain the code for the synthesis of a protein containing only a single amino acid. Addition of polyU resulted in the incorporation of phenylalanine alone into a protein resembling polyphenyl alanine. PolyU appears to function as a synthetic template or messenger RNA in this system".

28

HAR GOBIND KHORANA ALSO PLAYED A STELLAR ROLE IN CRACKING THE GENETIC CODE

Nirenberg's discovery that the artificial messenger RNA polyuracil, coded for the artificial protein polyphenyl alanine, attracted attention of Har Govind Khorana to this area, who was then in Madison, Wisconsin, U.S.A.

Har Gobind Khorana was born in a Hindu family in Raipur in the province of Punjab of India in 1922. This place is now in Pakistan. India was then undivided and under the British rule. His father worked for the government as a village agricultural taxation clerk. Har Gobind's task every morning was to fetch an ember to light the cooking fire at home. Smoke coming out of a house was his signal.

Khorana wrote in his autobiography, "Although poor, my father was dedicated to educating his children and we were practically the only literate family in the village inhabited by about 100 people. The first four years of my education were provided under a tree, a spot that was, in effect, the only school in the village". Quite often Har Gobind would be seen sitting on the steps of the village post office transcribing letters for the illiterate villagers.

He was good in studies and finished school and college earning the M.Sc. degree in 1945 aided by scholarships. Then he went to England on an India Government fellowship to do Ph.D. in Organic Chemistry. He received his PhD in 1948 from the University of Liverpool and then worked on alkaloid chemistry for a year in Eidgenössische Technische Hochschule in Zurich with Professor Vladimir Prolog. Subsequently he worked with Professor G. W. Kenner and Professor A. R. Todd in Cambridge. This is where he learned how to chemically join nucleotides.

In 1952 he was offered an independent position in the Organic Chemistry section of the University of British Columbia in Vancouver, Canada. There he chemically synthesized several molecules that were important for enzyme activity as cofactor, but he was still an Organic Chemist. In 1960 he was offered the position of co-director of the Institute of Enzyme Research at the University of Wisconsin in Madison, U.S.A.

Khorana , however, shifted his interest from enzyme research to cracking the genetic code, as Marshall Nirenberg announced in 1961 that polyuracil coded for polyphenyl alanine. He transformed himself into a Molecular Geneticist.

Har Gobind Khorana

Khorana utilized his knowledge of chemically synthesizing polynucleotides in constructing different heteropolymers of nucleotides (polymers of mixed nucleotides). For example, Khorana and coworkers found that the repeating dimer of U and C, *i.e.* UCUCUCUC…. coded for the polymer of alternating two amino acids, serine-leucine-serine-leucine-…, as UCU coded for serine and CUC coded for leucine. The repeating trimer of U, A and C, *i.e.* UACUACUACUAC......coded for three different polymers of single amino acids, i.e. polytyrosine, polyleucine and polythreonine. Depending on where the reading is initiated, UACUACUACUAC......can be read as UAC-UAC-UAC-UAC, or ACU-ACU-ACU-ACU, or CUA-CUA-CUA-CUA. The triplet UAC codes for tyrosine, the triplet ACU codes for threonine and the triplet CUA codes for leucine. Similarly Khorana's group chemically synthesized all other possible repeating dimers and repeating trimers and determined the polymers of amino acids produced. This approach enabled them to allocate the triplet codons for all the 20 amino acids.. Har Gobind Khorana was, however, not a member of the RNA Tie Club.

A former colleague of the University of Wisconsin had written, "Khorana was an early practitioner, and perhaps a founding father, of the field of Chemical Biology. He brought the power of chemical synthesis to bear on deciphering the genetic code, relying on different combinations of trinucleotides". In 1970 he moved to the Massachusetts Institute of Technology. There he chemically synthesized two complete genes and showed that these were biologically active.

Khorana then switched his fields and devoted himself in understanding the biochemistry of eye sight. He synthesized the gene for rhodopsin, a light sensitive protein involved in vision and studied the structure and function of rhodopsin. He pursued this study till his retirement in 1970.

Khorana was awarded the Louisa Gross Horwitz Prize, the Lasker Foundation Award, the Willard Gibbs Medal, the Gairdner Foundation Award, the Paul Kayser International Award and the Nobel Prize. He was elected as a Foreign Member of the Royal Society.

Khorana enjoyed hiking and would take long walks daily, often contemplating experiments while walking. He enjoyed listening to western classical music. His home was filled with paintings and books on science, art and philosophy. He was known for his modesty and humility and dislike for publicity.

Har Gobind Khorana died in 2011 at the age of 89.

29

SEVERO OCHOA ALSO PLAYED A ROLE IN CRACKING THE GENETIC CODE

The race for cracking the code was also joined by another American scientist named Severo Ochoa, who was then in the New York City. He had received the Nobel prize in 1959 for the discovery of the enzyme polynucleotide phosphorylase, that appeared then to synthesize RNA. Polynucleotide phosphorylase could be used for synthesizing polymers of ribonucleotides including triplets. Ochoa made use of this enzyme and also contributed the codes for some of the amino acids. Severo Ochoa, however, was not a member of the RNA Tie Club.

Severo Ochoa de Albornoz was born in 1905 in Luarca, a small mountain town in the district of Asturias in northern Spain. According to Spanish naming custom, the paternal last name Ochoa is mentioned first and the maternal last name de Albornoz is mentioned next. Severo's father, who was a lawyer and a businessman, died when Severo was only seven year old. Because of indifferent health of Severo's mother, the family moved to Malaga, a city with a warm climate on the Mediterranean coast.

Severo attended a private school in Malaga and then entered the Malaga College. He, wanted to pursue a career in Science, but found Mathematics rather tough. Hence he chose Biology and received his bachelor's degree in 1921. He then went to Madrid to study medicine and obtained the medical degree in 1929.

Severo Ochoa de Albornoz

Severo Ochoa had no intention of practicing medicine, so he went to do research as a post-doctoral fellow with Professor Otto Meyerhoff at the Kaiser-Wilhelm Institute in a suburb of Berlin. In 1931 he moved to London's National Institute for Medical Research and worked there with several enzymes. In 1933 he returned to the University of Madrid, where he was appointed a Lecturer in Physiology and Biochemistry. Within two years Ochoa moved again, this time to the newly established Institute of Medical Research as the Director of the section on Physiology.

In a few months, however, the Civil War broke out in Spain. Finding it difficult to continue his research work, Ochoa left Spain in 1936 and went back to Professor Otto Meyerhof at the Kaiser-Wilhelm Institute in Germany. Professor Meyerhof, however, being Jewish was then under considerable political and personal pressure, as the Nazis had come to power in Germany. In less than a year, he had to abandon his laboratory and flee to France. Before fleeing, Meyerhof arranged for Ochoa a six-month fellowship at the Marine Biological Laboratory in Plymouth, England. After six months, he found a position of Research Assistant in Oxford University.

As the second World War started in 1939, all laboratories had to work on projects related to war effort. Ochoa being a foreigner, could not be accommodated in this effort. So Ochoa had to move again, this time to the U.S.A. He found a job in the laboratory of Carl Ferdinand Cori in St. Louis. Cori was well known for his work on enzymes. After a year, Ochoa found an independent position in Bellevue Psychiatric Hospital under the New York University, where he could have graduate students and post-doctoral fellows working for him.

Ochoa remained in New York University till his retirement. He became an Assistant Professor in the Biochemistry department 1945, Professor in 1947 and Chairman of the Department in 1954. In 1955 he found an enzyme from a bacteria isolated from sewage, which could join ribonucleotides to form polyribonucleotides. He thought he had discovered the enzyme that converted ribonucleotides into RNA.

He was awarded the Nobel prize for this work in 1959. Ochoa retired at the age of 69 in1974, after serving as Chairman of the Biochemistry department of the New York University for 20 years.

It is said that Ochoa had fallen in love with two women, one was Carmen García Covián, whom he married, while the other was Enzymology. His love for both lasted throughout his life.

In 1975, the 70th anniversary of Severo Ochoa was celebrated by organizing a symposium in which many of his friends, colleagues and students participated. The proceedings were published in a book entitled, "Reflections in Biochemistry", the cover of which had a drawing by Salvador Dali, the famous artist and friend of Ochoa. In 1979 Ochoa wrote a memoir that he entitled "Pursuit of a Hobby", as he concluded "In my life Biochemistry has been my only and real hobby".

Besides the Nobel Prize, Ochoa was awarded the Carl Neuberg Medal, the Charles Leopold Mayer Prize, the U.S. National Medal of Science, the Medal of the Société de Chimie Biologique and the Medal of New York University.. He was one of the main sponsors of the Spanish Society of Biochemistry, the President of the Harvey Society, the American Society of Biological Chemistry, and the International Union of Biochemistry. He was member of the US National Academy of Sciences, of the American Philosophical Society, of the American Academy of Arts and Sciences. He was a Foreign Member of the Royal Society and of the Soviet Academy of Science. In 2011, the United States Postal Service issued a stamp honouring him, as part of the *American Scientists* collection. An asteroid has been named after him, Asteroid 117435 Severochoa.

Ochoa returned to his native Spain in 1985 to serve as a Professor of Biology at the University Autonoma in Madrid. Ochoa's love for his country was amply reciprocated. The Center of Molecular Biology in Madrid, Spain was renamed as the Centro de Biología Molecular Severo Ochoa (CBMSO). South of Madrid a hospital has been named after him, Hospital Severo Ochoa. Many streets in many cities of Spain have been named after him. Many restaurants that Ochoa visited display his photograph.

Severo Ochoa died in Madrid in 1993 at the age of 88..

30

THE FOLLOWING ARE THE GENETIC CODES FOR THE DIFFERENT AMINO ACIDS

Phenylalanine:	UUU, UUC
Lysine:	AAA, AAG
Proline:	CCC, CCA, CCU, CCG
Leucine:	UUA, UUG, CUU, CUC, CUA, CUG
Isoleucine:	AUU, AUC, AUA
Valine:	GUU, GUC, GUA, GUG
Methionine:	AUG
Serine:	UCU, UCC, UCA, UCG, AGU, AGC
Threonine:	ACC, ACG, ACU, ACA
Alanine:	GCU, GCC, GCA, GCG
Tyrosine:	UAU, UAC
Histidine:	CAU, CAC
Glutamine:	CAA, CAG

Asparagine:	AAU, AAC
Lysine:	AAA, AAG
Aspartate:	GAU, GAC
Glutamate:	GAA, GAG
Cysteine:	UGU, UGC
Arginine:	CGU, CGC, CGA, CGG, AGA, AGG
Serine:	AGU, AGC
Glycine:	GGU, GGC, GGA, GGG
Chain Termination:	UAA, UAG, UGA

The codes UAA, UAG or UGA would signal termination of the synthesis of protein. It may be noted that all amino acids except methionine have more than one triplet code, but no code specifies more than one amino acid. A particular code is specific for a particular amino acid. So in the language of DNA there are many synonymous words, but there is no ambiguous word.

31

FRANCIS CRICK'S ADAPTOR HYPOTHESIS TO EXPLAIN HOW THE PROTEIN SYNTHESIZING MACHINERY TRANSLATED THE BASES OF THE TRIPLET CODES IN THE MESSENGER RNA INTO THE AMINO ACIDS

Once the genetic codes for all the 20 different amino acids were known, the question that remained to be answered was how exactly the protein synthesizing machinery translated the bases of the triplet codes in the messenger RNA into the amino acids. It was none else than Francis Crick from Cambridge, England, who around mid 1955 came forward and circulated a note to the members of the

RNA Tie Club entitled, "On degenerate templates and the adaptor hypothesis". He asserted that it was structurally impossible for any triplet of bases to directly act as a template for and recognise a particular amino acid. So he enunciated 'the adaptor hypothesis'—the presence of a class of molecule that would act like an adaptor in an intermediate step of protein synthesis. Crick continued, "The adaptor could consist of perhaps a single chain of RNA"

To understand Crick's logic, we have to go back to the description of amino acids in Chapter 11, "An interesting element of all the protein amino acids is that these compounds comprise the same basic feature, a simple linear structure with the same ends, we would call these as (N) end and (C) end. The difference between one amino acid and the other is in the side chain attached to this basic linear structure". Crick felt that the possibility is very remote that the code comprising three bases would recognize the side chain of an amino acid.

What he meant was that the three bases of the code in the messenger RNA would recognise three bases present in the adaptor RNA molecule if they are complementary, (as defined by Watson and Crick's base pairing rule (Chapter 19), because hydrogen bonds can occur between the two sets of ribonucleotides. An example would be the code UUU in messenger RNA being recognised by the triplet AAA in the adaptor RNA, as they can base pair. AAA can be termed as the anticode of code UUU.

Crick also hypothesized that the adaptor molecule containing the anticode would be joined to the amino acid that is specific for the said code by an enzyme that would be specific for the said adaptor and the amino acid. If UUU is the code, the amino acid would be phenylalanine and the adaptor RNA would contain the anticode AAA. A specific enzyme would join the amino acid phenylalanine with its cognate adaptor RNA that would contain the anticode AAA.

Francis Crick had later described his notion about the adaptor in his lecture entitled "On Protein Synthesis", that he presented at the 1957 annual meeting of the Society of Experimental Biology, which was subsequently published in 1958. Crick had no experimental evidence, the adaptor hypothesis was based solely on his intellectual exercise.

32

PAUL ZAMECNIK OBTAINED THE FIRST CLUE OF THE INTERMEDIATE STEP IN PROTEIN SYNTHESIS FOR TRANSLATING THE TRIPLET OF BASES IN THE MESSENGER RNA INTO THE AMINO ACIDS

For obtaining experimental evidence in favour of Crick's adaptor hypothesis, we travel across the Atlantic Ocean from Cambridge, England to the University Hospitals in Cleveland, U.S.A. We discover that the act really started much earlier, in 1938. Paul Charles Zamecnik, after his M.D. and residency from Harvard Medical School, was doing his internship in the University

Hospitals in Cleveland. An obese woman under his care had died within a week of hospitalization. Autopsy revealed too much fat, but very little protein in her tissues. Zamecnik inferred that her protein synthesizing machinery was not functioning, he also realized that he knew nothing about protein synthesis. He talked to a number of biochemists, but they all drew blank. That was the beginning, Zamecnik became mentally occupied in understanding how proteins were synthesized.

Paul Charles Zamecnik was born in 1912 in Cleveland, Ohio, U.S.A. His father was a banker and his grandfather was a musician, composer and conductor. He attended Dartmouth College and then Harvard Medical College, receiving his MD degree in 1936. He did his internship in the University Hospitals in Cleveland. He was awarded the Finney-Howell Fellowship and a Moseley Traveling Fellowship to work with Dr. Kai Linderstrom-Lang in Copenhagen. He had to return to U.S.A. prematurely in 1940, as the Second World War started and the Germans occupied Denmark. In 1942 he joined the Department of Medicine at the Harvard Medical School in Boston as an Instructor.

The breakthrough came in 1953. Zamecnik was then full Professor in the Harvard Medical School. He had his own laboratory and a research team comprising Mahlon Hoagland and Elizabeth Keller. They were successful in preparing an extract from rat liver that was free of any intact cell, but was capable of synthesizing proteins from amino acids. Apparently there were enough endogenous messenger RNA in the rat liver extract, in contrast to the *Escherichia coli* extract of Marshall Nirenberg (Chapter 27). They used radioactive amino acids as trackers to monitor their conversion into proteins.

In the beginning of the experiment, only the amino acids were radioactive, but the proteins were not radioactive. As the experiment progressed more and more proteins were becoming radioactive. In 1956, they were successful in detecting an intermediate moiety.

Some RNA molecules were acquiring radioactivity even before any protein was acquiring any radioactivity. The acquired radioactivity of the RNA molecules must have been the result of the radioactive amino acids getting hooked on to the RNA molecules forming an RNA – amino acid complex. When they isolated the RNA molecules that had acquired radioactive amino acids and added fresh rat liver extract to it, the radioactive amino acids got transferred to the proteins. They inferred that for formation of proteins, all amino acids must first get hooked on to some specific kind of RNA molecules, only then the amino acids can get transferred to the proteins being formed. These RNA molecules were just acting as carrier in the transfer of the amino acids. Obviously, these specific kind of RNA molecules must be recognising the triplet code of the specific amino acid they were acting as carrier of the amino acid. They named the particular species of RNA molecules as transfer RNA (tRNA in abbreviated form) and on analysis found these to be only about 80 bases long.

Paul Charles Zamecnik

The transfer RNAs thus could be the adaptors that Francis Crick had hypothesized (Chapter 31). It has been mentioned in chapter 25 that some soluble RNAs containing about 80 bases are present in a cell of *Escherichia coli*. These happen to be the transfer RNA molecules of *Escherichia coli*.

Zamecnik and coworkers also discovered that the platform on which protein synthesis occurred was the ribosomes. As has been mentioned in chapter 25, in all cells there are some particulate entities named ribosomes, which are complexes of a few species of RNA and a large number of proteins. The species of RNA present in ribosomes are larger than and different from the transfer RNAs.

Some enzymes must be involved in joining the tRNA molecules with its cognate amino acid molecules. These enzymes were subsequently purified by Zamecnik and co-workers.. Zamecnik's transfer RNA matched with the hypothetical characteristics of Crick's adaptor.

Amino acid joined to a tRNA was named aminoacyl tRNA and the enzyme which did the joining was named aminoacyl tRNA synthetase. Examples are leucyl tRNA synthetase, prolyl tRNA synthetase, alanyl tRNA synthetase, valyl tRNA synthetase, etc. It may be mentioned here that Paul Zamecnik was not a member of the RNA Tie Club.

Zamecnik finally became Professor of Medicine in the Harvard Medical School and served until retiring as the Collis P. Huntington Professor of

Oncologic Medicine, Emeritus in 1979. He continued his research activity at the Worcester Foundation for Biomedical Research and then at the Massachusetts General Hospital in Boston until several weeks before his death.

According to Thoru Pederson, Professor of Cell Biology at the University of Massachusetts, "Dr. Zamecnik's work advanced the study of protein synthesis, taking it from a purely metabolic process into proper Biochemistry for the first time". Francis Crick said, "Zamecnick's discovery was one of the highest moments of his life in science". The Lasker jury, while bestowing on him the Albert Lasker Award said, "Zamecnik's identification of transfer RNA is a landmark in the history of Science. In this one experiment Biochemistry and Genetics became fused into the new science of Molecular Biology".

Paul Zamecnik was a very pleasant and well mannered person. Herman Kalckar, the great Biochemist of that era, likened him with the French actor, Maurice Chevalier. In case Zamecnik disagreed with someone, he would never say it bluntly. He would say pleasantly, "Well, as a Harvard Medical School Dean liked to say: 'We can agree to disagree." He would be referred to as the perfect gentleman by many scientists of his time. Zamecnik's love for literature was also well known, he would recite classical literature and poetry eloquently.

Paul Zamecnik was awarded the National Medal of Science, the American Society for Biochemistry and Molecular Biology Merck Award, American Cancer Society National Award. He was a member of the U.S. National Academy of Sciences, American Academy of Arts and Sciences, American Society of Biological Chemistry, American Association for Cancer Research, Association of American Physicians, Danish Academy of Sciences and Letters and the American Philosophical Society.

Paul Zamecnik died of cancer in 2009 at the age of 96.

33

ROBERT WILLIAM HOLLEY DETERMINED THE BASE SEQUENCE OF A TRANSFER RNA AND REVEALED THE PRESENCE OF AN ANTICODE IN IT

Paul Zamecnik discovered that an amino acid must first be hooked on to its cognate transfer RNA before it could be incorporated into proteins, transfer RNAs thus have a very important role to play in protein synthesis. But what does a transfer RNA look like?

Does it really harbour the anticodon which is complementary to the triplet codon specific for the amino acid which is carried by the transfer RNA?

We travel from Boston to Cornell University in Ithaca in the neighbouring New York state. It is 1954, we meet there Associate Professor of Organic

Chemistry, Dr. Robert William Holley. Holley believed that to understand what a transfer RNA looks like and to find out whether it harbours the anticodon predicted by Francis Crick, one must determine its base sequence.

Robert William Holley was born in 1922 in Urbana, Illinois, U.S.A. Both his parents were school teachers. He graduated from High School in 1938 and obtained his B.A. degree in Chemistry 1942 from the University of Illinois. He obtained the Ph.D. degree in Organic Chemistry in 1947 from Cornell University.

He worked as post-doctoral fellow for a year and then joined the faculty of Cornell University as Assistant Professor of Organic Chemistry. He was promoted to Associate Professorship in 1950. In 1953, Holley went on a year's sabbatical leave to work with James Bonner in the California Institute of Technology at Pasadena. Bonner was a famous Plant Biochemist, his book on Plant Biochemistry was more of a textbook. Holley worked on protein synthesis with Bonner and the idea of specifically working on transfer RNA took root there. After coming back to Cornell University from California Institute of Technology, Holley initiated his efforts on determination of the base sequence of a transfer RNA.

Holley realised that isolating a particular transfer RNA from all the RNA present in any cell would be almost like picking up a needle from a hay stack. So he did not choose any plant system for his work. He chose baker's yeast as the source material which was commercially available in plenty. He did not have to worry about growing the cells or plants.

In 1958, Holley moved to the Plant, Soil and Nutrition Laboratory of the United States Department of Agriculture, Agricultural Research Service of U.S.A. at the Cornell University campus. Though it was a laboratory that had a mandate of working particularly for Plant nutrition, the Director, W. H. Allaway allowed Holley to continue working on yeast transfer RNA.

It has been mentioned in chapter 25 that a cell contains three varieties RNA in the ribosomes, which are about 3,000 bases, about 1,500 bases and about 100 bases long. In addition, there would be the messenger RNAs (chapter 25) whose average size would be about 900 bases. The transfer RNAs are about 80 bases long. Holley's first job was to develop a method for the separation of all the transfer RNAs from other RNAs present in the cell and their purification. .Since transfer RNA molecules would be specific for each amino acid, there would be at least 20 different transfer RNA molecules for the 20 different amino acids. Holley subsequently developed the method of isolation of an individual transfer RNA from the mixture of all transfer RNAs. Holley focused his attention on a particular transfer RNA, the transfer RNA that served as the carrier of the amino acid alanine for incorporation into proteins (alanyl transfer RNA).

Holley and his co-workers obtained 200 gm of total mixed transfer RNA from 150kg of Baker's yeast. They then obtained from the mixed transfer RNA, 1 gm of pure alanyl transfer RNA. Next the task of determination of the base sequence began and by the end of 1964 Holley's team achieved their goal of sequencing all the bases of alanyl transfer RNA. The whole project took almost 10 years, but the achievement was indeed colossal. The base sequence of a complete nucleic has been unravelled for the first time.

Tracy Sonneborn, Professor of Biology at the Indiana University, U.S.A. called it a "marvelous achievement" and Maxine Singer, Chief of the Laboratory of Biochemistry at the National Cancer Institute, Bethesda, U.S.A. described it as a "formidable job."

Holley proposed a clover leaf model for the transfer RNA (Figure 35). About middle of the length of the alanyl transfer RNA molecule, there is a loop and at the tip of the loop is located the base triplet CGI, where I stands for inosine, a purine very similar to adenine. In DNA, adenine base pairs with thymine, but since uracil has replaced thymine in RNA, adenine base pairs with uracil in RNA. The base pairing characteristics of inosine is very similar to the base pairing characteristics of adenine, hence inosine base pairs with uracil. The triplet CGI then can base pair with the triplet GCU and therefore can serve as the anticodon of the codon triplet GCU of alanine.

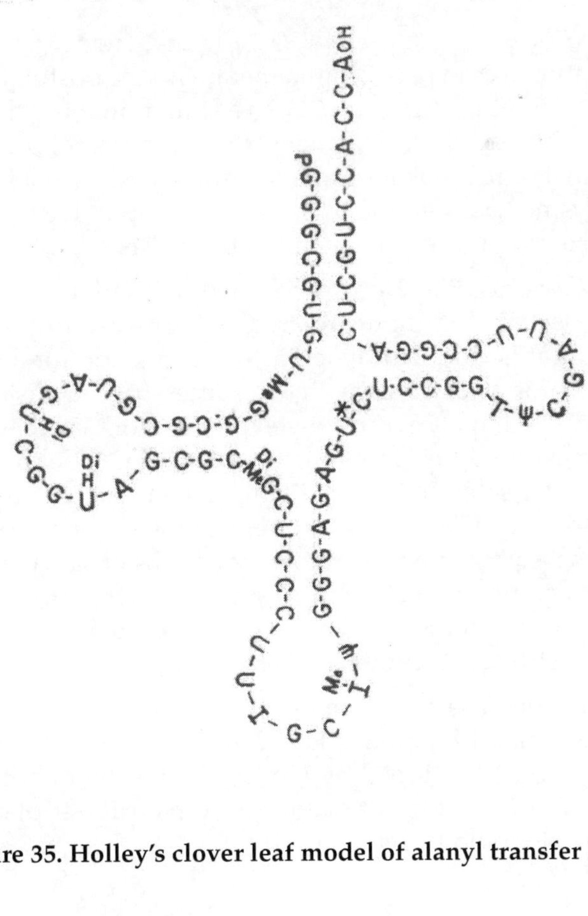

Figure 35. Holley's clover leaf model of alanyl transfer RNA.

Robert William Holley

This observation was in perfect agreement with the postulate of the adapter hypothesis of Francis Crick (Chapter 31). In a similar manner the transfer RNA that is specific for the codon UUU of phenylalanine, has later been found to have the triplet AAA in its anticodon loop and the transfer RNA that is specific for the codon AAA of lysine has been found to have the triplet UUU in its anticodon loop. Robert Holley also was not a member of the RNA Tie Club.

Holley was awarded the Nobel prize for this achievement. He had said during the Nobel lecture , "Without minimizing the pleasure of receiving awards and prizes, I think it is true that the greatest satisfaction for a scientist comes from carrying a major piece of research to a successful conclusion." He did not forget to thank W.H.Allaway, who allowed him to work on yeast transfer RNA in a laboratory mandated for plant nutrition. He said, "Special thanks are due Dr. W. H. Allaway, Director of the U. S. Plant, Soil and Nutrition Laboratory, for his confidence and encouragement throughout the sequence determination." "The moral is to have good administrators - much easier said than done." At the end of his Nobel acceptance lecture he said in a lighter vein, "That then is our story of the alanine transfer RNA. It all followed quite naturally from taking a sabbatical leave. I strongly recommend sabbatical leaves."

Subsequently, the base sequence of all other transfer RNAs that acted as the carrier of the other 19 amino acids were determined by other scientists following the methods developed by Holley. About the middle of the length of all these transfer RNAs, the anticodon specific for the triplet codons of the other 19 amino acids could be recognised.

Holley later changed his field and worked on the control of cell division in mammalian cells.

In addition to the Nobel Prize, Holley received the Albert Lasker Award, the Distinguished Service Award of the U. S. Department of Agriculture and the U. S. Steel Foundation Award in Molecular Biology of the National Academy of Sciences. He was elected member of the National Academy of Sciences, the American Academy of Arts and Sciences, the American Association for the Advancement of Science, the American Society of Biological Chemists and the American Chemical Society. Robert W. Holley died of cancer in1993 at the age of 71 years.

34

SYNTHESIS OF PROTEINS

As we have travelled from the Pasteur Institute in Paris to the Cornell University in Ithaca visiting on the way the Enzyme Institute in Madison, the New York University in New York City, the Cambridge University in Cambridge, England and the Harvard Medical School in Boston, we saw François Jacob **and** Jacques Monod mooting the concept of messenger RNA as carrier of message from DNA (Chapter 24), Marshall Nirenberg, Har Gobind Khorana and Severo Ochoa discovering the base triplets in messenger RNA that coded for each amino acids during synthesis of proteins (Chapters 27, 28 and 29), Francis Crick enunciating the adaptor hypothesis postulating the involvement of an intermediary in protein synthesis that would contain nucleotides and would be able to pair with the triplet code (Chapter 31), Paul Zamecnik demonstrating that transfer RNAs are the adaptors (Chapter 32) and Robert Holley determining the base sequence of a transfer RNA and predicting its structure. This leads us to the final act, how exactly proteins are synthesized.

As has been mentioned in chapters 25 and 32, the platform on which synthesis of proteins takes place are some cytoplasmic particles named ribosomes. A schematic representation of the process of protein synthesis appears in Figure 36 .

Step 1 in protein synthesis is the attachment of a ribosome to the end of the messenger RNA, where the triplet codon AUG for initiation of protein synthesis happens to be present. Then this first triplet codon AUG in the messenger RNA would align itself with a site in the ribosome named the "A" site or site

I. An aminoacyl tRNA (aminoacyl transfer RNA), whose anticodon (UAC) is complementary to the first triplet codon (AUG) in the messenger RNA, would attach to the "A" site or site I and its anticodon (UAC) would hydrogen bond with the first triplet codon (AUG) in the messenger RNA.

In step 2, the messenger RNA along with the hydrogen bonded aminoacyl tRNA would move, so that the first codon-anticodon complex is aligned to another site on the ribosome called the "P" site or site II, leaving the "A" site on the ribosome vacant.

In step 3, the "A" site or site I on the ribosome would be aligned with the second triplet codon of the messenger RNA. A second aminoacyl tRNA, whose anticodon is complementary to the second triplet codon in the messenger RNA, would attach to the "A" site or site I of the ribosome and its anticodon would hydrogen bond with the second triplet codon in the messenger RNA. Now an amide bond would be formed between the carboxyl end of the first amino acid and the amino end of the second amino acid, forming a dipeptide

In step 4, the messenger RNA would move the second time so that the second codon-anticodon complex is aligned to the "P" site or site II, leaving the "A" site or site I of the ribosome vacant. The bond between the amino acid and the tRNA of the first aminoacyl tRNA would then be snapped and so also the hydrogen bonds between the first triplet codon and the anticodon of the tRNA. The released tRNA would then go to the pool of all free tRNAs. The second tRNA at the "P" site or the site II of the ribosome would have a dipeptide (Figure 12) linked to it instead of a single amino acid (Figure 11).

In step 5, a third aminoacyl tRNA, whose anticodon is complementary to the third codon in the messenger RNA, would attach to the "A" site or site I and its anticodon would hydrogen bond with the third triplet codon in the messenger RNA. An amide bond would be formed between the carboxyl end of the dipeptide linked to the second tRNA and the amino end of the third amino acid carried by the third aminoacyl tRNA, forming a tripeptide.

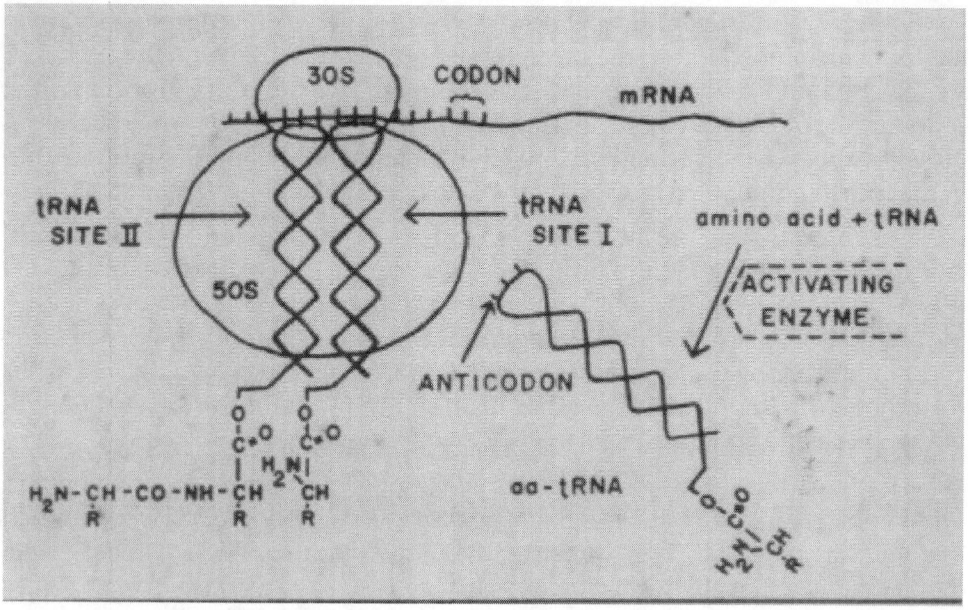

Legend of Figure 36. Schematic representation of the process of protein synthesis in the colon bacteria *Escherichia coli.* *The Figure has been adapted from the Figure 12 of the paper by H. K. Das, Avram Goldstein and Lee C. Kanner (1966) Inhibition by chloramphenicol of the growth of nascent protein chains in Escherichia coli. Mol.Pharmacol. 2: 158-170. The ribosome, which is the platform on which proteins are synthesized, comprises two subunits, 50S and 30S. The first transfer RNA that initiated the synthesis protein is not shown in the Figure . Its anticodon was complementary to the first triplet codon and it has left the scene after delivering its amino acid. The messenger RNA has moved and the second triplet is now at the P site or the site II of the ribosome. The second tRNA (transfer RNA) (the structure shown is a simplified one) can be seen at the "P" site or the site II carrying a dipeptide linked to it, which is the result of peptide bond formation between the first amino acid and the second amino acid. A third aminoacyl tRNA, whose anticodon is complementary to the third codon in the messenger RNA, is attached to the "A" site or site I and its anticodon has hydrogen bonded with the third triplet codon in the messenger RNA. Those days amino acyl tRNA synthetase used to be called amino acid activating enzyme. In the Figure, the fourth aminoacyl tRNA after being formed from the amino acid and the cognate tRNA, is waiting for its turn to move to the site A or site I. The configurations of the tRNA molecules are more complex than shown in the Figure.*

In step 6, the messenger RNA would move the third time so that the third codon-anticodon complex is aligned to the "P" site or site II, leaving the "A" site or site I vacant. The bond between the amino acid and the tRNA of the second aminoacyl tRNA would then be snapped and so also the hydrogen bonds between the second triplet codon and the anticodon of the second tRNA. The second tRNA would then go to the pool of all free tRNAs. The third tRNA at the "P" site or site II of the ribosome would have a tripeptide linked to it instead of a dipeptide.

In step 7, a fourth aminoacyl tRNA, whose anticodon is complementary to the fourth triplet codon in the messenger RNA, would then attach to the "A" site or site I and its anticodon would hydrogen bond with the fourth triplet codon in the messenger RNA. An amide bond would then be formed between the carboxyl end of the tripeptide linked to the third tRNA and the amino group of the fourth amino acid carried by the fourth aminoacyl tRNA.

In subsequent steps, all the triplet codons of the messenger RNA would be read sequentially by the anticodon regions of the respective aminoacyl tRNAs using Watson-Crick base complementarity rule and correctly line up the amino acids. Amide linkages among the carboxyl end of the oligopeptide and amino end of the next amino acid carried by the next aminoacyl tRNA would be formed one after the other resulting in the synthesis of a full length polypeptide or a protein.

That was the final episode of the story of Demystification of Gene and The People who made it Possible. The story of the people who discovered that heredity is the attribute of genes, that genes control chemical reactions in living systems, that chemical reactions are catalysed by enzymes, that enzymes are proteins, that proteins are strings of amino acids, that the genes comprise DNA, that the sequence of deoxyribonucleotides in DNA is transcribed into the sequence of ribonucleotides in RNA and that the triplets of ribonucleotides in messenger RNA code for the amino acids in proteins. There is no mystery any more.

35

RECAPITULATION OF THE STORY IN BRIEF

It all began in 1894 B.C. in the court of the new king Sumuabum of the city of Babylon. The new king had expressed his displeasure that the local dates, though bigger were less sweet. The wise old man looked for groves outside Babylon which produced sweeter dates (though smaller) and collected pollens from the flowers from their male plants. He dusted these pollens on to the flowers of female plants in Babylon. When the fruits appeared later, some of the large dates were indeed sweeter. The wise old man had no inkling as to what exactly the tiny pollens could be carrying.

Four thousand years later Gregor Johann Mendel was dusting pollens from short garden pea plants on to the flowers of tall garden pea plants after removing its stamen. Interestingly, all the progeny plants had tall stems, no plant with short stem or stem of intermediate size. Whatever the pollens were carrying was not reflected in the height of the pea plants. When flowers of these plants were allowed to get pollinated by their own pollens, about one-fourth of the plants that grew were short stemmed. So, the information carried by the pollens from the short plants was not really lost, but not expressed in the first generation. The information in pollens was indelible. Mendel termed it hereditary "factor", but he had no idea as to what comprised a hereditary "factor" or how it acted.

Around five decades later Wilhelm Ludwig Johannsen carried out experiments similar to that of Gregor Johann Mendel, but with princess bean and obtained similar results. In 1909 he coined the term "gene" to describe the hereditary "factor". What really was the nature of "gene" or how it influenced the characters of the plants was a mystery to Johannsen.

About a decade earlier to Johannsen's crossing experiments with bean plants, physician Archibald Edward Garrod came across an infant, whose urine had stained the diaper brownish-black. This ailment was known as Alkaptonuria. Three years later the mother of the infant came again with her second child, who also was suffering from Alkaptonuria. The woman had married her first cousin. She and her husband had the same grandfather, who had this ailment, but she and her husband were free of it. Garrod found other similar cases. He was convinced that Alkaptonuria was inherited by a child from both parents, though they have it in a silent or symptomless form. So it was a case of indelible defective hereditary information. Garrod had no clue about what was the nature of the information that was defective. He, however, guessed that the defective hereditary information had resulted in the absence of a chemical reaction, as hereditary information must be controlling chemical reactions in the body.

More than 40 years ahead of Garrod's time, the Sweedish doctor turned chemist, Jöns Jacob Berzelius discovered chemical catalysis. A catalyst is a substance that "mediates a chemical reaction or enhances the rate of a chemical reaction without itself undergoing any permanent chemical change". Many chemical reactions need high temperature to occur, but a catalyst would often enable the reaction to take place at a lower temperature. Berzelius had guessed that chemical reactions in a living object might also be mediated by catalysts, as these reactions go on at the ambient temperature or at 37°C in warm blooded animals. He could not guess though what the nature of the catalysts in living objects would be. Genes were thought to be controlling the chemical reactions in living beings, could genes be controlling the catalysts that mediate the chemical reactions in living systems? Nobody knew how.

Around the middle of the nineteenth century Louis Pasteur was studying how grape juice was converted into wine by the microscopic living organism yeast. He concluded that a "vital force" present in the yeast cells was doing it and it "was an act correlated with the life and organization of the yeast cells".. Pasteur could not conceive that catalysts, as predicted by Berzelius, could be involved in the conversion process.

In1877, Wilhelm Kühne opined against Pasteur's view that a "vital force" in yeast cells could be responsible for conversion of grape juice into wine. Instead, he supported Berzelius's idea that biological catalysts were doing the job. He coined the term "enzymes" for the biological catalysts. "Enzymes" were catalysing chemical reactions in yeasts and all other living systems. Though it was guessed that the enzymes were chemical entities, the nature of the enzymes or what the enzymes were composed of was a mystery. Also mysterious was how the genes were controlling the enzymes.

Two decades later Eduard Buchner demonstrated that wine could be produced by replacing living yeast cells by an extract of crushed yeast cells that contained no living yeast in it. This nullified Pasteur's notion of the necessity of a "vital force" that was "correlated with the life and organization of the yeast cells" and supported Kuhne's view that biological catalysts or "enzymes" were involved in the process. Buchner gave the name "zymase" to the particular enzyme(s) responsible for wine production. Again Buchner did not know anything about the nature of "zymase" or its relationship with the gene.

James Batcheller Sumner had the conviction that there was nothing mysterious about the enzymes and decided to embark on the task of unravelling the nature of enzymes. He realized that he must have an enzyme in very pure state before he attempted to characterize it. Crystals were considered to be very pure and Sumner tried to isolate the enzyme urease in crystalline form from jack bean meal. After several years of relentless efforts he finally was successful in 1926. This was the first time that any one obtained an enzyme in crystalline form. Using specific chemical tests, Sumner discovered that the crystals of the enzyme urease were nothing but a protein. Other enzymes were subsequently crystallized by Sumner himself and other scientists by following Sumner's method. Those enzymes were also found to be proteins.

But what are proteins? Jerardus Johannes Mulder first described "protein" in 1838 being the main constituent of albumin in the white portion of hen egg. Franz Hofmeister and Emil Fischer independently found out in 1902 that proteins were linear array of many different amino acids.

There are 20 different amino acids in proteins. All amino acids have an amino group, that is basic and a carboxyl group, that is acidic, at the two ends. The difference between one amino acid and the other is in the side chain attached to this main linear structure. The carboxyl end of amino acid (1) would react with amino end of amino acid (2) forming a link. The carboxyl end of amino acid (2) would react with amino end of amino acid (3). The carboxyl end of amino acid (3) would react with amino end of amino acid (4) and so on and on resulting in a chain of many amino acids comprising the protein. The sequence of amino acids in a protein, not just what amino acids are contained, determines the property of the protein, hence the property of the enzyme.

As mentioned earlier, Archibald Edward Garrod had guessed that the defective hereditary information, hence a defective gene, had resulted in the absence of a chemical reaction, thus causing Alkaptonuria. Since chemical reactions in living beings are catalyzed by enzymes, the defective gene might have produced a defective enzyme. Around the end of 1930s, George Wells Beadle and Edward Lawrie Tatum wanted to artificially create defects in genes. They chose *Neurospora*, the pink mould that grows on stale bread as the experimental object and exposed these to X-rays. Normally, *Neurospora* could be grown in a medium containing the simple sugar glucose, common salt, a few other minerals and biotin, a B vitamin. Many of the *Neurospora* died, some that survived exposure to X-ray, could no longer grow on the simple medium. Beadle and Tatum found out that these happened because some enzymes that catalysed chemical reactions which produced various compounds from the simple growth medium were permanently inactivated by X-ray exposure. They could correlate these inactivated enzymes with specific genes that became defective because of X-ray exposure. This was the first experimental evidence that specific genes produced specific enzymes., "one gene one enzyme", as Beadle and Tatum stated.

Frederick Griffith observed in 1928 that when killed infectious slimy form (S form) of the bacteria *Diplococcus pneumoniae* were mixed with live non-infectious rough form (R form) of the same bacteria and injected into mice, the mice got killed . The blood of the dead mice was full of live slimy form (S form) of the bacteria. So something from the dead slimy form (S form) of bacteria had transformed the non-infectious rough form (R form) into the infectious slimy form (S form). Could it be the gene for the enzyme which catalysed the chemical reaction that produced slime? Griffith did not seem to worry about it. He was possibly more concerned with patterns of pneumoniae epidemiology.

In 1944, Oswald Theodore Avery purified the transforming principle from the slimy form (S form) of the bacteria *Diplococcus pneumoniae.* Chemical tests revealed that it was DNA. When the substance was treated with an enzyme that degraded DNA, the transforming activity was lost, but when it was treated with an enzyme that degraded proteins, the transforming activity was unaffected. When the DNA was mixed with live non-infectious rough form (R form) of *Diplococcus pneumoniae* and mice were injected with the mixture, the mice got killed and the blood of the dead mice was full of S form of the bacteria. So Mendel's hereditary factor or Johannsen's gene was nothing but DNA.

Johannes Friedrich Miescher isolated DNA from human pus cells in 1869. He named it "nuclein". It contained phosphorus, but no sulphur, while all proteins contained sulphur, but no phosphorus. The substance was present in sperms of quite a large number of animals.

DNA is a long linear molecule, the individual units being deoxyribonucleotides. A deoxyribonucleotide comprises the sugar deoxyribose attached to a purine or pyrimidine base at one end and a phosphate group at the other end. The deoxyribose moiety of one deoxyribonucleotide is linked to the deoxyribose moiety of the adjacent deoxyribonucleotide through the phosphate group. The purine base of DNA is either adenine (A) or guanine (G), while the pyrimidine base is either thymine (T) or cytosine (C). The size of the purine bases, G and A are about the same, while the size of the pyrimidine bases thymine (T) and cytosine (C) are about the same. The size of the purine bases G and A is, however, almost double the size of the pyrimidine bases C and T.

Erwin Chargaff found in 1950 that the proportion of a particular base relative to all the four bases was different in DNA from man, ox, yeast and avian tuberculosis bacilli. This was in accordance with the view that DNA was the genetic material. He also observed that the proportion of adenine (A) and thymine (T) was about the same, while the proportion of guanine (G) and cytosine (C) was also about the same in the DNA of all these species.

In a bid to understand the three dimensional structure of DNA, Rosalind Franklin of Kings College, London was successful in determining its X-ray diffraction pattern in 1951, Her famous Photo 51 was really decisive. John Desmond Bernal, the famous X-ray crystallographer had said that "Her photographs were among the most beautiful X-ray photographs of any substance ever taken. Their excellence was the fruit of extreme care in preparation and mounting of the specimens as well as in the taking of the photographs." The X-like pattern formed by the layer lines immediately suggested to Franklin a closely packed helical structure of DNA having the phosphate chains near the outside. The intensity of the reflection at the meridian was, however, too strong for a single helix. The structure "indicated that the molecule was in two matching parts, running in opposite directions, evidence convincing enough that DNA had a double helical structure". She also realized "that the way the component nucleotides or bases on each strand were connected meant that the two strands were complementary, enabling the molecule to replicate." Franklin was never in a haste to publish or build models without enough irrefutable experimental data.

James Watson and Francis Crick of Cambridge University, England were trying to construct a molecular model of DNA, but they did not have any experimental data of their own. Watson attended the lecture in King's College that Rosalind Franklin delivered in November 1951, where she had presented her X-ray crystallographic data and enumerated her inferences. On January 30, 1953 Maurice Wilkins, another scientist of Kings College, had shown Photo 51 to Watson without caring for the simple ethics of seeking permission from Rosalind Franklin. As Watson recalled later, "that when he saw the photo – which was far clearer than any other he had seen – 'my mouth fell open and my pulse began to race.'" Watson went back to Cambridge and described to Crick what he saw in photo 51. The annual work report of the King's College submitted in December, 1952 contained all the detailed crystallographic calculations done by Rosalind Franklin to arrive at her conclusion that DNA had a double helical structure. . Max Perutz, Crick's Ph.D. thesis adviser, had a copy of that report and around the middle of February 1953, he passed on that report to Crick. Watson and Crick knew now that DNA had two strands, but not three as they had earlier thought. Also the phosphates were outside and the bases were inside, not *vice versa,* in contrast to what they had conceived earlier. Next Watson and Crick took the cue from Erwin Chargaff's finding that irrespective of the source of DNA, the proportion of adenine was the same as the proportion of thymine, and the proportion of guanine was the same as the proportion cytosine. So they constructed a model double helix placing adenine in one strand against thymine in the other strand and placing guanine in one strand against cytosine in the other strand. Watson and Crick published their model of DNA in the scientific journal Nature of April 25,1953. In their article there was no clear acknowledgement that they had obtained crucial clues for building their model from Franklin's photo 51and her crystallographic calculations.

In their first article on DNA structure, Watson and Crick wrote, "It has not escaped our notice that the specific pairing we have postulated immediately suggests a possible copying mechanism for the genetic material". On this point also Rosalind Franklin had mentioned in her report, "that the way the component nucleotides or bases on each strand were connected meant that the two strands were complementary, enabling the molecule to replicate." Watson and Crick elaborated in their second article, "We imagine that prior to duplication the hydrogen bonds are broken, and the two chains unwind and separate. Each chain then acts as a template for the formation on to itself of a new companion chain, so that eventually we shall have two pairs of chains, where we only had one before."

The next question was whether DNA could directly guide the synthesis of a protein. The problem is that in higher forms of life like animals and plants, DNA has been found located inside the nucleus, while protein biosynthesis takes place in the cytoplasm of the cell. May be, there is a group of molecules that originate inside the nucleus and has the information of base sequence of DNA, which travel to the cytoplasm and direct the amino acid sequence of the proteins. Indeed, in all cells there is a group of molecules similar to DNA which is called RNA. RNA also is a long linear molecule, but RNA has ribose instead of deoxyribose contained by DNA. The purine base of RNA is also either adenine (A) or guanine (G), while the pyrimidine base is either uracil (U) or cytosine (C). So while DNA has thymine, RNA has uracil instead.

In 1933 Jean Brachet, found out that both DNA and RNA were present in all cells, irrespective of its source, but while DNA was mostly in the nucleus, RNA was in the cytoplasm. Brachet observed that a specific RNA fraction first appeared in the nucleus and then got transferred into the cytoplasm. When he treated the cells with the enzyme ribonuclease, which degrades RNA, protein synthesis was inhibited, but when he removed the nucleus, protein synthesis continued in the cytoplasm of the cell for some days before ultimately stopping.. This was the decisive indication that DNA did not directly guide protein synthesis, but probably did it through a species of RNA that was not very stable.

Francis Crick enunciated the Central Dogma of Molecular Biology in 1957, which said, DNA → RNA → protein. More precisely, it stated that genetic information stored in DNA can flow into protein through RNA, but it can not flow directly from DNA to protein. As before, he did not do any laboratory experiment for enunciating the Dogma.

François Jacob and Jacques Monod mooted the concept of messenger RNA in 1961.

The protein they studied synthesis of in the bacteria *Escherichia coli*, was the enzyme β-galactosidase, which split the sugar lactose. When they fed the bacteria with lactose, it synthesized lot of β-galactosidase. Then they took a strain of the bacteria that needed uracil for its growth and fed both uracil and lactose to it. Again lot of β-galactosidase was synthesized. If they washed the bacteria and fed lactose but no uracil, no β-galactosidase was synthesized. The synthesis of β-galactosidase could be resumed if uracil was added back to the growth medium. So synthesis of an enzyme, which is a protein, needed uracil, which is present in RNA but not in protein. Jacob and Monod hypothesized that synthesis of a specific type of RNA was essential for the synthesis of the enzyme β-galactosidase. The named this type of hypothetical RNA as messenger RNA. Francois Jacob then teamed up with Sydney Brenner and Mathew Meselson and was successful in actually isolating total messenger RNA of *Escherichia coli*, a heterogeneous RNA species that spread between sizes of 1,500 bases and 80 bases.

Since there are 20 different amino acids in a protein and there are 4 different bases in DNA, minimum of 3 bases would be needed to code for an amino acid. This is because if any two bases would code for one amino acid, then $4^2 = 16$ amino acids would only be coded. That three bases of DNA indeed coded for an amino acid, was verified by Sydney Brenner, Francis Crick, and co-workers in the 1950s. They drew their inference from the observation that deletion of one base pair, or two successive one base pairs (achieved by two successive treatments with proflavine) from a particular gene of the virus T4 rendered its protein product inactive, but three successive one base pair deletions restored the activity in some cases, at least partially.

Marshall Nirenberg demonstrated in 1961 that an artificial messenger RNA containing polyuracil (an RNA chain of all uracils) directed the synthesis of polyphenyl alanine (a protein chain of all phenyl alanines). Since the code was a triplet, UUU was the code for phenylalanine. He used the lysate of the colon bacteria *Escherichia coli* and that was free of intact cells, as source of the protein synthesizing machinery and used radioactively labelled amino acids to track their incorporation into proteins. Nirenberg also obtained synthetic ribohomopolymer of adenine (an RNA chain of all adenines) and ribohomopolymer of cytosine (an RNA chain of all cytosines) and found out that they acted as messenger RNA for polylysine (a protein containing all lysines) and polyproline (a protein containing all prolines) respectively. Thus the codes for the amino acids lysine and proline were AAA and CCC respectively. Subsequently, he found out all the codons for each of the 20 amino acids present in proteins by synthesizing the codons biochemically using an enzyme-based approach.

Har Gobind Khorana also played a stellar role in cracking of the genetic code. He chemically synthesized polynucleotides and found that the repeating dimer of U and C, *i.e.* UCUCUCUC.... coded for the polymer of alternating two amino acids, serine-leucine-serine-leucine-..., as UCU coded for serine and CUC coded for leucine. Similarly his group chemically synthesized all other possible repeating dimers and repeating trimers and determined the polymers of amino acids produced. This approach enabled them to allocate the triplet codons for all the 20 amino acids..

Severo Ochoa also played a role in cracking the genetic code. He had discovered the enzyme polynucleotide phosphorylase, which could be used for synthesizing polymers of ribonucleotides including triplets in the laboratory. Ochoa made use of this enzyme and also contributed the codes for some of the amino acids.

One problem remained to be solved. How are the triplet codes in messenger RNA recognized by the amino acids during protein synthesis? Francis Crick asserted that structurally it was impossible for an amino acid to directly recognize ribonucleotides. He enunciated in 1955 'the adaptor hypothesis'—the presence of a class of molecule that would act like an adaptor in an intermediate step of protein synthesis. The adaptor molecules would contain complementary nucleotides, which would be able to recognize the triplet code in the messenger RNA by base pairing according to Watson and Crick base pairing rule. For example, the code UUU would be recognized by the triplet AAA, which can be termed as the anticode. Crick also hypothesized that the adaptor molecule would be joined to the specific amino acid by an enzyme that would be specific for the said adaptor and the amino acid. If UUU is the code, the amino acid would be phenylalanine and the adaptor would contain the anticode AAA. A specific enzyme would join the amino acid phenylalanine with its specific adaptor that would contain the anticode AAA.

Paul Zamecnik was successful in experimentally detecting an intermediate moiety of protein synthesis in 1956.. He was using cell free extract of rat liver as the system for protein synthesis by radioactive amino acids. Some RNA molecules were acquiring radioactivity even before any protein was becoming radioactive. When they isolated the RNA molecules that had acquired radioactivity and added fresh rat liver extract to it, the radioactivity in the RNA molecules got transferred to the proteins. He named the particular species of RNA molecules as transfer RNA (tRNA in abbreviated form) and on analysis found these to be only about 80 nucleotides long. The enzymes involved in joining the tRNA molecules with its cognate amino acid molecules were subsequently purified by Zamecnik and co-workers and named aminoacyl tRNA synthetases. Zamecnik's transfer RNA matched with the hypothetical characteristics of Crick's adaptor.

Robert William Holley developed a method for the purification of all the transfer RNAs from the cell lysate of Baker's yeast and also the method of isolation of individual transfer RNAs from the mixture of transfer RNAs. He purified alanyl tRNA, the transfer RNA that acted as the carrier of the amino acid alanine and determined the sequence of bases present in it. About the middle of the length of the transfer RNA, he detected the presence of the triplet CGI, where I stands for inosine, a purine very similar to adenine with base pairing characteristics of adenine. The triplet CGI can serve as the anticodon of the codon triplet GCU of alanine, as CGI is complementary to the codon GCU according to Watson and Crick's base pairing rule. The requirement of Crick's hypothetical adaptor was thus fully met. Subsequently, the base sequence of all other transfer RNAs were determined by other scientists following the methods developed by Holley. All of them have the specific anticodon.

The platform on which synthesis of proteins takes place are some cytoplasmic particles named ribosomes. Proteins are synthesized in a stepwise fashion. In step 1, the messenger RNA attaches on to the ribosomes and the first triplet codon aligns with the 'A' site of the ribosome. The cognate transfer RNA carrying the specific amino acid recognizes the codon through complementarity of the anticodon in it and attaches to the 'A' site. In step 2, the messenger RNA along with the aminoacyl transfer RNA moves such that the first triplet codon is aligned to the 'P' site of the ribosome, vacating the 'A' site. The second triplet codon of the messenger RNA is now aligned with the vacant 'A' site. In step 3, the second amino acyl transfer RNA, which has an anticodon complementary to the second triplet codon recognizes the second triplet codon and occupies the 'A' site. A peptide bond is formed between the first amino acid and the second amino acid. The bond between the first amino acid and its cognate transfer RNA now breaks. The first transfer RNA detaches from the 'P' site and goes to the transfer RNA pool. The messenger RNA now moves with the second transfer RNA and the dipeptide attached to it, such that the second triplet codon has now shifted to the 'P' site vacating the 'A' site. In step 4, the third triplet codon is aligned to the vacant 'A' site. The third amino acyl transfer RNA, which has an anticodon complementary to the third triplet codon recognizes the third triplet codon and occupies the 'A' site. A peptide bond is next formed between the dipeptide and the third amino acid, forming a tripeptide. The bond between the second amino acid and its cognate transfer RNA now breaks. The second transfer RNA detaches from the 'P' site and goes to the transfer RNA pool.. The messenger RNA again moves with the third transfer RNA and the tripeptide attached to it, such that the third triplet codon has now shifted to the 'P' site vacating the 'A' site. In step 5, the fourth triplet codon is aligned to the vacant 'A' site. The fourth amino acyl transfer RNA, which has an anticodon complementary to the fourth triplet codon recognizes the fourth triplet codon and occupies the 'A' site. A peptide bond is now formed between the tripeptide and the fourth amino acid, forming a tetrapeptide and so it goes on finally forming a long polypeptide or protein.

That was the recapitulation of the story of Demystification of Gene and The People who made it Possible.

Source Locations

Following are the sources from which information has been gathered:

en.wikipedia.org; www.sciencedirect.com; www.cell.com; www.researchgate.net; www.ncbi.nim.nih.gov; www.yourgenome.org; www.nobelprize.org; www.pasteur.fr; www.britannica.com; www.dnaftb.org; www.nasonline.org; www.crick.ac.uk; www.famousscientists.org; www.sciencehistory.org; www.oxfordreference.com; www.biography.yourdictionary.com; www.profiles.nim.nih.gov; www.biography.com; www.nature.com; www.royalsocietypublishing.org; www.jewishvirtuallibrary.org; www.encyclopedia.com; www.genetics.org; www.hindustantimes.com; www.whatisbiotechnology.org; www.khanacademy.com; www.nytimes.com; www.genomenewsnetwork.com; www.acs.org; www.pnas.org; www.genome.gov; www.biologydictionary.net; www.newsharvard.edu; www.academictree.org; www.sciencemag.org; www.mayoclinicproceedings.org; www.ars.usda.gov; www.newscientist.com; www.nationalmedals.org; www.chemindigest.com.; www.president.uchicago.edu

Photographs of all scientists have been reproduced from Wikipedia.